Practices and Procedures of Industrial Electrical Design

Practices and Procedures of Industrial Electrical Design

L. B. Roe
Managing Director, L. B. Roe (Consultants) Ltd., and American Marketing Systems
Los Angeles, California, and Blackpool, England
Member, Institute of Electrical and Electronics Engineers

McGRAW-HILL BOOK COMPANY
**New York   St. Louis   San Francisco   Düsseldorf   Johannesburg
Kuala Lumpur   London   Mexico   Montreal   New Delhi
Panama   Rio de Janeiro   Singapore   Sydney   Toronto**

Library of Congress Cataloging in Publication Data

Roe, L   B
  Practices and procedures of industrial electrical design.

  1. Electric engineering. I. Title.
TK145.R74   621.3   70-168754
ISBN 0-07-053390-3

Copyright © 1972 by McGraw-Hill, Inc. All rights reserved. Printed in the United States of America. No part of this publication may be reproduced, stored in a retrieval system, or transmitted, in any form or by any means, electronic, mechanical, photocopying, recording, or otherwise, without the prior written permission of the publisher.

1 2 3 4 5 6 7 8 9 0 KPKP 7 6 5 4 3 2

The editors for this book were Tyler G. Hicks and Stanley E. Redka, the designer was Naomi Auerbach, and its production was supervised by George E. Oechsner. It was set in Caledonia by The Maple Press Company.

It was printed and bound by The Kingsport Press.

*Dedicated to my parents
Mr. and Mrs. G. B. Roe and B. J. W.*

1724866

Contents

Preface *v*

PART 1   Electrical Design Information

1. Definitions and Standards   3

   1.1   National Electrical Code   3
   1.2   Definitions   4
   1.3   Standards   4
   1.4   Symbols   4
   1.5   Notes   5
   1.6   Revisions   5
   1.7   Signatures and Approvals   6
   1.8   Inspection and Permits   6
   1.9   Extras   6
   1.10   Specifications   7

2. The Basic Electric System   8

   2.1   The System and Primary Components   8
   2.2   Measurement and Control   8

Contents

- 2.3 Equipment 9
- 2.4 The Generator 9
- 2.5 The Circuit Breaker 10
- 2.6 The Stepup Transformer 10
- 2.7 The Circuit-breaker Location 11
- 2.8 The Transmission Line 11
- 2.9 The Stepdown Transformer 12
- 2.10 The Distribution System 12
- 2.11 The Distribution Transformer 12
- 2.12 Protection and Control 13
- 2.13 The Substation 13

3. Job Preparation 15

- 3.1 The Single Line 15
- 3.2 The Hold 16
- 3.3 The Plot Plan 17
- 3.4 The Drawing List 19
- 3.5 The Job File 20

4. Motors and Exciter Wiring Calculations 22

- 4.1 Basic Requirements [NEC 430-1(a)] 22
- 4.2 Squirrel Cage (Three-phase Motor) 22
- 4.3 Three-phase Induction-motor Calculations 22
- 4.4 Multiple Motors 25
- 4.5 Future Motor Installations 27
- 4.6 Breaker Trip Setting 28
- 4.7 Code Compliance 29
- 4.8 A Logic Analysis 29
- 4.9 The Wound-rotor Motor 31
- 4.10 Secondary Conductors 32
- 4.11 Conductor Sizing 33
- 4.12 The Single-phase Motor 34
- 4.13 Direct-current (DC) Motors 36
- 4.14 Synchronous Motors 37
- 4.15 DC Rotor Field 38
- 4.16 The Exciter 40
- 4.17 Size Evaluation and Short-circuit Ratio 41
- 4.18 Conductor Size 42

5. Generators and Variable-speed Drives 44

- 5.1 Generators 44
- 5.2 The AC Generator—Three-phase 44
- 5.3 Generator Ratings 45
- 5.4 Power Factor 46
- 5.5 Power 46
- 5.6 Generator Output 47
- 5.7 Generator Conductors 48

5.8   The Exciter   49
5.9   Sample Calculations   50
5.10 Cost Evaluation   50
5.11 DC Generators   51
5.12 The DC Generator Exciter   51
5.13 Variable-speed Drives   52
5.14 Horsepower and Load   54
5.15 Motor-starting Characteristics   56

6. Transformers 57

6.1   Transformer Function   57
6.2   Transformer Principles   57
6.3   Transformer Selection   58
6.4   The Autotransformer   59
6.5   Transformer Rating   59
6.6   Autotransformer Limitations   61
6.7   Power Transformers   61
6.8   The Single-phase Transformer   61
6.9   Distribution Transformers   62
6.10 Control Power Transformers   62
6.11 Potential Transformers   62
6.12 Isolating Transformer   62
6.13 Current Transformers   63
6.14 The Zigzag Transformer   63
6.15 Constant-current Transformer   65
6.16 Transformer Overcurrent Protection   66
6.17 Differential Protection   66
6.18 Polarity   68
6.19 Subtractive and Additive Polarity   68
6.20 Transformer Connections   69

7. Power and Energy 81

7.1   Power   81
7.2   Power Measurement   81
7.3   Apparent Power   82
7.4   Active Power   82
7.5   Phase Angle   82
7.6   Time Period   83
7.7   Real Power   83
7.8   Reactive Power   83
7.9   Power Addition   84
7.10 Vector Addition   84
7.11 Ammeter Reading   84
7.12 Power Factor   85
7.13 Unity Power Factor   86
7.14 Lagging Power Factor   86
7.15 Oversize Motors   86
7.16 Oversize Exciter   86

7.17 Leading Power Factor   87
7.18 Power-factor Correction   87
7.19 Leading-var Generators   88
7.20 Power-factor-correction Calculations   88
7.21 Plant Power-factor Correction   89
7.22 System Capacity Release   91
7.23 Switched Capacitors   92
7.24 System Analysis and Capacitor Location   93
7.25 Power-factor Calculations of Mixed Loads   93
7.26 Capacitors on Distribution Lines   95
7.27 Capacitors Installed at Transformers   96
7.28 Motor and Switched Capacitor Installation   96
7.29 Calculation of Capacity   98

8. System Study 99

8.1 Scope   99
8.2 The Single Line   99
8.3 Input Parameters   100
8.4 Reason for Study   100
8.5 Study Layout   101
8.6 Degree of Complexity   101
8.7 Protective-device Coordination   101
8.8 Fuses versus Circuit Breakers   102
8.9 The Circuit Breaker   102
8.10 The Recloser   103
8.11 The Sectionalizer   103
8.12 The Evaluation   103
8.13 Relays—Analog Type   103
8.14 Relays—Digital Binary Type   103

9. The Short-circuit Study 105

9.1 Short Circuits   105
9.2 Breaker Interrupting Capacity (IC)   106
9.3 Bolted Short Circuit   106
9.4 Short-circuit Current   107
9.5 The Short-circuit Study   107
9.6 Fault Location   108
9.7 Equipment Impedance   108
9.8 Impedance Parameters   108
9.9 Percentage Values   109
9.10 Degree of Accuracy   109
9.11 Equipment Rating—Transformer   110
9.12 Base KVA   111
9.13 Rotating Equipment   111
9.14 Equipment Reactances   114
9.15 Momentary Rating   114
9.16 Interrupting Rating   114
9.17 Impedance Diagram   115

9.18	Star Delta Conversion	118
9.19	Short-circuit Magnitude	119
9.20	Assymmetrical Current	120
9.21	Rules of Thumb	120
9.22	Motor Control Center	121
9.23	Summary	121

10. Instrumentation and Control Circuits 122

10.1	Project Conception	122
10.2	Flowsheet	122
10.3	P&ID	123
10.4	Analog Instruments	123
10.5	Digital Instrumentation	123
10.6	Protective Relays	123
10.7	Control Relays	124
10.8	Logic Arrangement	124
10.9	Control Contacts	124
10.10	Relay Variations	125
10.11	Selector Switches	125
10.12	Limit Switches	126
10.13	Control-circuit Parameters	126
10.14	Program Construction and Analysis	126
10.15	The Rewrite	127
10.16	Logic Connectives	129
10.17	The Circuit Layout	130
10.18	STOP/START Circuit	130
10.19	HAND/OFF/AUTO	131
10.20	Control Switches	132
10.21	Logic Variables	132
10.22	Control-switch Option	133
10.23	Auxiliary Relay	133
10.24	Multiple STOP/START	135
10.25	STOP/START and HAND/OFF/AUTO	135
10.26	Undervoltage Release	136
10.27	Summary	136

11. Circuit Logic 138

11.1	Mathematical Logic	138
11.2	Circuit Logic	139
11.3	"Permissive" or "Hindrance" Solution?	139
11.4	The Logical AND	140
11.5	The Logical OR	141
11.6	De Morgan's Law	142
11.7	The Exclusive OR	144
11.8	The AND-OR Circuits	145
11.9	Networks	146
11.10	Redundancy and Absorption	148

12. Wiring and Connection Diagrams 152

12.1 Control Diagram 152
12.2 Schematic Diagram and Numbering 152
12.3 Wiring Diagram 153
12.4 Connection Diagram 153
12.5 Wiring and Connection Diagrams Combined 154
12.6 Airway Connection Method 155
12.7 Color Coding 156

13. Grounding 157

13.1 Definition 157
13.2 System and Equipment Grounding 158
13.3 System Grounding 158
13.4 Delta System 158
13.5 Delta System Ground 159
13.6 The Star System 160
13.7 Solid, Resistance, Reactance Grounding 160
13.8 Ground Location 160
13.9 Low-voltage Systems 161
13.10 Identified Conductor 161
13.11 Grounding Conductor 161
13.12 Equipment Grounding 161
13.13 Enclosure Grounding 162
13.14 Nonelectric Equipment Grounding 162
13.15 Portable Equipment Grounding 162
13.16 Structural Grounding 162
13.17 Lightning Conductors 162
13.18 Bonding 163
13.19 Grounding Source 163
13.20 Water Pipe Bonding 163
13.21 Structural Steel Grounding 163
13.22 "Made" Electrodes 164
13.23 Ground Wells 164
13.24 Ground Resistance 164
13.25 Ground Testing 165

14. Lighting 166

14.1 The Art of Lighting 166
14.2 Lighting Design 166
14.3 Lighting Calculations 167
14.4 Lumen Method 167
14.5 Validity of Lighting Calculations 168
14.6 Rule of Thumb 169
14.7 Fixture Spacing 170
14.8 Louvers and Diffusers 171
14.9 High-bay Lighting 171
14.10 Fixture Selection 171

14.11 Candlepower Distribution Curve   172
14.12 Point-to-point Calculation   175
14.13 Circuiting   178
14.14 Floodlighting   178
14.15 Summary   179

15. Checking 181

15.1 Drawing Issue   181
15.2 Conditions of Issue   181
15.3 Drawing Check   182
15.4 Color Code Checking   182
15.5 Checking Philosophy   182
15.6 Drafting Technique   183

PART 2   Simplified Design Mathematics

16. Number Manipulation 187

16.1 The Scalar   187
16.2 The Equality   187
16.3 The Sum   188
16.4 The Difference   188
16.5 The Product   188
16.6 Division   189
16.7 Unknown Quantities   190
16.8 Cross Multiplication   192
16.9 The Exponent   195
16.10 The Radical   197
16.11 Powers of Ten   199

17. Trigonometry and the Radius Vector 201

17.1 The Right Triangle   201
17.2 The Coordinate Axis   202
17.3 The Radius Vector   202
17.4 Trigonometric Functions   203
17.5 Pythagorean Theorem   204
17.6 Common Functions   205
17.7 Quadrants   205
17.8 Angles Greater than 360°   205
17.9 Functions of Angles Greater than 90°   206

18. Vector Manipulation 209

18.1 Vector Representation   209
18.2 Location of a Point   209
18.3 Polar Form   210
18.4 Rectangular Form   211

18.5 Cartesian Form 212
18.6 Polar, Rectangular, and Cartesian Transposition 214
18.7 Operator j 214
18.8 Vector Operations 216
18.9 Vector Manipulation 217

19. Mathematical Logic 220

19.1 Principles of Logic 220
19.2 The Proposition 221
19.3 Negation 222
19.4 The Logical Sum 223
19.5 The Logical Product 224
19.6 The Logical Equality 224
19.7 The Implication 227
19.8 Modus Ponens 229
19.9 De Morgan's Law 230
19.10 The Exclusive OR 230
19.11 Application of Mathematical Logic 232

Appendix 235

Table A.1 Ampacities 236
Table A.2 Conduit Fill 237
Table A.3 Properties of Conductors 238
Table A.4 Circuit-breaker Types 239
Table A.5 Circuit-breaker Application Ratings 240
Table A.6 Motor-branch-circuit Data—460-volt Three-phase AC 242
Table A.7 Transformer Data 243
Table A.8 NEMA Enclosure Classification 244
Table A.9 Temperature Conversion 245
Table A.10 Three-phase Motors Full Load Currents 246
Table A.11 Natural Trigonometric Functions 247
Table A.12 American Standard Device Function Numbers 248
Table A.13 Selected Symbols 249
Table A.14 Electrical Formulas 252
Table A.15 The Greek Alphabet 252
Table A.16 Electrical and Magnetic Nomenclature 253
Table A.17 Rule-of-thumb Approximations 254

Index 255

Preface

This book is written for the whole of the engineering profession, for practicing engineers as well as for students. Thus it is a reference book containing facts which all members of the field may find useful. However, it is aimed at the workhorse of the industry: the designer.

Because the book is an electrical book, naturally it is written with the electrical designer in mind. It assumes that the designer has some previous experience, but that there will always be problems arising with which he is not completely familiar.

Depending upon the reader's background, then, the contents of the book may occasionally seem "out of order" or possibly too advanced or too elementary. The subject matter was selected and arranged in accordance with two criteria:

1. Material should be presented that will make the designer's task easier and speedier.
2. The information should be presented in the order of requirement as a design progresses.

This is why power-factor correction, short-circuit studies, and checking occur toward the middle or end of Part I rather than as fundamentals at the beginning.

It was decided to place the reference mathematics as Part 2 of the book. This decision was made because it is basically an electrical book, not a mathematics book. It is essential, however, that the mathematics part be perused *first*. Unless the reader is familiar with classical logic, trigonometry, and vectors, his understanding of the electrical portion will be limited.

Project engineers from other fields such as civil, mechanical, and chemical engineering should have no difficulty in understanding matters of electrical design after reading this book or a specific chapter of it.

Because the designer seems to be the forgotten individual as far as practical textbooks are concerned, the author would welcome any suggestions as to future material to include. As revisions are introduced the designers can eventually have their own design standard.

Thanks to Charles J. Helin, P.E., Past President, Board of Registration for Professional Engineers, State of California, for his review of the manuscript and his helpful suggestions.

Illustrations are by John Smets, of Huntington Park, California.

L. B. Roe

**Practices and Procedures
of Industrial
Electrical Design**

Part One

Electrical Design Information

Chapter One

Definitions and Standards

1.1 National Electrical Code

The National Electrical Code is a publication of the National Fire Protection Association. It is one of a number of codes covering various fields. The Code is advisory in intent and has no legal standing unless it is specified in individual contracts that it be the legal guide for minimum safety requirements. In this book we will use the National Electrical Code (NFPA-70) as the standard code. Each individual state, city, county, etc., may have its own individual electrical code. More often than not, the individual codes are based on the National Electrical Code with some modifications to suit the local requirements. Sometimes these are more stringent than the National Electrical Code. Therefore, when designing an electrical installation in a particular area, contact the electrical inspection department in that area and obtain a copy of the local code; otherwise obtain in writing the procedure used by the local inspection department for evaluating minimum requirements of an installation.

The purpose of the National Electrical Code is to set certain minimum standards necessary for the protection of persons and property against the hazards of misuse of electrical equipment, faulty equipment, or faulty design and installation. Equipment that meets the minimum requirements outlined in the Code will not necessarily have the most efficient design—this is not the intent of the Code. It is assumed by the Code authority that the design is carried out and supervised by competent individuals trained in electrical design. Any questionable section of the Code should be submitted, in writing, to the local enforcement agency for a ruling before going ahead with the design associated with that section. This eliminates the possibility of the contractor installing equipment in a manner which may violate that particular code. The National Electrical Code is revised periodically; therefore it should be unnecessary to point out that old copies should not be used.

1.2 Definitions

One of the major requirements for the electrical designer is an up-to-date copy of the National Electrical Code. For the purpose of definitions we will use the guide in Article 100 of the NFPA-70 Code. Because a copy of NFPA-70 is an essential, we will not repeat all the definitions it contains. Any conflict between definitions given in this book and those in NFPA-70 will be explained later, in the appropriate place in the text.

1.3 Standards

The term *standard* refers to routine design procedures set forth by an authority. These procedures are usually in the form of a loose-leaf book or a set of drawings. The standards usually cover routine parts of the normal design process. The purpose of the standards is to obtain better conformity and at the same time to cut down on the design time. These standards will vary from company to company, each company having its own ideas on what a standard should be. Thus a standard may imply standardization only within a particular organization. When reporting to a new company ask for the copy of its standards.

1.4 Symbols

The symbols used throughout the industry are as varied as are the standards. When reporting to a new company it is essential to inquire if a standard set of symbols is used by the company before beginning any design work. It is embarrassing to complete a section of design and find that the symbols differ from those used in the remainder of the design.

In this text we will use the symbols given in the Appendix. These symbols have been selected with two points in mind: first, the symbol is an abstract picture intending to convey graphically the equipment it represents; second, simplicity is essential because of drafting time involved. The resulting symbol then represents a compromise between these two considerations.

1.5 Notes

There are two basic types of notes: general notes and sheet notes. The general notes are "standard" type notes and apply to any drawing in a particular design package. These general notes are usually written on the first drawing in the design package, and as previously stated, they should apply to any drawing in that package. The sheet notes are notes associated with a single sheet only. A general note may also be a sheet note. When writing notes, care must be taken with the wording; remember that the person reading them may be from another part of the country where colloquial expressions have a different meaning. The length of the note is a compromise between brevity to conserve drafting time and enough wording to make the message unmistakably clear. The engineer or constructor in the field may spend considerable time and expense trying to interpret a note whose meaning is ambiguous.

1.6 Revisions

A revision of a drawing may involve either an alteration of a previous part of the design or an addition to the design. The change may be very minor or it may consist of a complete new drawing. In either case a record of this must be maintained and new copies must be issued to the parties holding previous drawings. When making revisions, there are certain considerations to be taken into account: (1) How many prints are required to replace all previous copies of the drawing? (2) Is the change significant and a necessity? (3) Does it affect purchasing, ordering, and delivery of material? (4) Will the revised drawing arrive in time for the field crew to make the necessary change? (5) Are all the approval signatures available? These five considerations are all obvious and should require no further explanation. One word of caution: When a revision is issued it should be accompanied by a strongly worded note to the resident in charge of construction that all previous copies of the drawing should be immediately taken from the field crew and replaced with the revised version. This sounds like a simple and obvious request; however, in actual practice the field crew have a habit of marking and making notations on their copy of the print. They then retain the void print and sometimes keep working from the void print on the assumption that the last revision did not affect their particular portion of the construction. This is not good practice; all void prints should be removed from the construction area and placed in a stick file in the construction office for reference and record purposes only.

1.7 Signatures and Approval

When a cashier's check is submitted to a bank, it is a rarity for the bank to honor the check without the correct signature. When a drawing is issued for construction, it is signed by the necessary people with the authority to approve construction as outlined on the drawing. This construction may cost thousands or even millions of dollars in labor and material, and yet sometimes more thought goes into signing the check for the mortgage payment than into the approval of a drawing. We do not advise an attitude of stubborn resistance when one must proceed with construction with an unsigned drawing. Rather, we would advise a diplomatic approach pointing out the oversight and requesting a revised drawing correcting the omission or a covering letter authorizing the use of the unsigned issue of the drawing. Remember, like the cashier's check, a drawing is not valid until approved with the correct signatures. And, again like the cashier's check, if it is honored without the correct signature there is no recourse in the event of errors.

1.8 Inspection and Permits

Usually both a permit and an inspection are required for any installation. The normal procedure in this case is to include in the design specification or general notes (if a specification is not required) the authorization for the contractor to complete all necessary requirements and arrangements to obtain the permits and inspections required by the state and local authorities. The constructor should also be warned that it is his responsibility to comply with all the state and local codes and regulations so that a final inspection does not present any expensive or time-consuming problems.

1.9 Extras

The term *extra* refers to work beyond the scope of the work outlined within the contractual agreement. The possibility of a project being completed within the scope of the contract is remote. With this in mind, extras must be considered. Within the design specification the procedure for approval of extras should be outlined. The same principles apply as in the conditions outlined in the section on signatures and approvals. A constructor doing work beyond the contract agreement without valid approval has no recourse for payment.

Many extras are carried out with no more than a verbal assertion that if they were necessary for the completion of the project then approval (after the fact) would be given. This type of agreement is common and is probably a valid contract if it meets the requirements for a contract, i.e., capacity of parties, lawful object, good consideration, and mutual consent of both parties, and if it meets the legal requirements for written contracts. Certain types of contracts must be in writing to be valid. The verbal type of agreement is not good; however, sometimes it is necessary. When this is the case the considerations should be weighed very carefully and this type of agreement should be used only when recourse in the form of future business or some other form of protection is available. When in doubt insist on giving or obtaining written approval before proceeding with the work (depending on whether you have construction or design responsibility).

1.10 Specifications

The *job specification* is an outline of the scope and procedures to be followed in the design and/or construction of a particular project. It is issued by the client in the form of a request to design a job in a particular way. When the design is completed a specification is issued by the design company to the construction company outlining the degree of responsibility and methods, procedures, codes, etc., that the constructors should be aware of. Individual specifications may be issued to vendor companies for equipment and material. In all cases no legal recourse is available if the specification did not outline the correct requirements. This is of course excluding the condition "with intent to defraud." When writing specifications a review of mathematical logic may establish the correct frame of mind for stating exactly the requirements intended. The use of mandatory and advisory grammatical connectives such as "shall" and "should" or "will" should be clearly defined. It is hard to compromise between the confusing terminology of the lawyer and the simplicity required for construction or design instructions. Experience in the various fields of the project is necessary to be able to hold a completely objective and realistic view and keep a specification within reasonable bounds.

Chapter Two

The Basic Electric System

2.1   The System and Primary Components

Electricity is simply another form of energy. It is only useful, however, when it can be controlled and transmitted. There are three basic requirements for the electric system: production, transmission, and utilization. The production of electricity requires a source of energy, such as water, steam, gas, fossil fuels, atomic energy, etc. These are then used to provide a "prime mover" with power. Attached to the "prime mover" is a generator. The generator produces electricity at a relatively low voltage. To prepare the electricity for transmission over long distances, it is fed into a device known as a transformer. The voltage is then raised by transformer action to approximately 1,000 volts for every mile of transmission. For 150 miles the voltage would probably be a standard 169 kv (169 kilovolts). When the fringe of the local area is reached, the electricity is once more fed into a transformer, which then reverses the process and lowers the voltage for local distribution.

2.2   Measurement and Control

The quantity of electricity used is measured by meters and other instruments. The amount delivered and the stability of the system are controlled by regulating devices, protective relays, and circuit breakers.

2.3 Equipment

Generally, all "transmitting" electric equipment will fall into one of the preceding categories. The actual size of the equipment will vary with the amount of voltage and power it is required to handle. For example, a circuit breaker can be as small as a cigarette package, or it can be large enough to walk around in. In either case, it serves the same basic function. To ensure a comprehensive understanding of the function of a particular piece of equipment, we will introduce the basic system. No attempt will be made to explain the design of a specific piece of equipment or the limits of its use. This will be covered in a later chapter. We will explain the general function of the equipment and introduce the correct terminology. Some terms tend to be ambiguous. Therefore, for the purpose of this text, we will establish the generic name and the specific duty of the equipment.

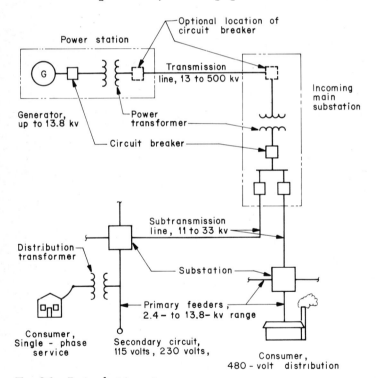

Fig. 2.1 Basic electric system.

2.4 The Generator

The *generator* is a device used to transform mechanical energy into electrical energy. The electricity may be in the form of *direct current*, *alternating single-phase current,* or *alternating polyphase current.* The selected frequency of the alternating current can be optional. The standard frequency in the United States and Canada is 60 cycles per second (cps) for normal utility systems. In aircraft and missile systems, varying frequencies are used. For example, 400 cps is used in some aircraft systems. This high frequency allows a reduction in the weight of components. The rating of ac generators is given in kva output at rated voltage and power factor. The dc generators are rated in kw output at rated voltage. kva is simply an abbreviation for kilovolt-amperes or va 10^{-3} (kilo meaning a thousand). The same applies to kw (this is an abbreviation for kilowatts). The standard generator used for conventional utility electric systems is always a three-phase unit.

2.5 The Circuit Breaker

The conductors leaving the generator terminate in a *circuit breaker.* The function of this piece of equipment is to prevent excessive currents in the system beyond the generator. At the same time, it can be made to serve as a protective device to prevent overloading of the generator. Circuit breakers come in many forms, shapes, and sizes. There are air circuit breakers, vacuum circuit breakers, oil circuit breakers, power circuit breakers, and plastic case circuit breakers. These are varied in current rating, voltage rating, short-circuit rating, and shock rating (mechanical shock). The methods of tripping the circuit breakers are also quite varied. Each manufacturer has particular features in his design which he feels may make it better than another manufacturer's design. Sometimes these features become standard and are common to all circuit breakers. With all these differences, all circuit breakers have one thing in common: They disconnect a circuit from the supply when the current in the system exceeds predetermined limits. The circuit breaker may also serve double duty as a disconnect device or switch, being operated manually instead of automatically.

2.6 The Stepup Transformer

We now begin our sequence of transmitting power from one location to another. We have, up to now—beginning with the generator—terminated in a circuit breaker. From here, we feed into a transformer. The purpose of the transformer is to raise the voltage from the rated voltage of the generator to a new voltage level, suitable for transmitting over long distances. This new voltage level is determined by evaluating a number of conditions. The voltage selected is dependent on installation cost, distance, operating expense, and also present and future voltages of other systems in the vicinity.

The types of transformers used for feeding transmission lines are usually custom built to meet the utility-company requirements. They may be three-phase autotransformers[1] with approximately a 2:1 ratio of transformation, or three-phase two-winding transformers with higher ratios of transformation. In some cases, three single-phase transformers are used and connected in star or delta to provide a three-phase system of transformation.

2.7 The Circuit-breaker Location

We will again introduce the circuit breaker at this point. As we are considering a logical sequence of major equipment, we must point out that apparent power is the product of voltage and current. For a given quantity of apparent power, we can consider low voltage and high current or high voltage and low current. Depending on the particular values in each case, selection of a circuit breaker must be made. The current values in the low-voltage system may be excessive for a circuit breaker, whereas in the high-voltage system, the current and voltage levels may be well within normal limits for standard circuit breakers. Therefore, we may find it necessary to locate the circuit breaker on the high-voltage side of the transformer. In either case, the same protection is offered to the system. There may also be other factors (such as the type of system the generator and transformer are being connected to) which will govern the location of the circuit breaker. The main point to consider is that there is a choice of locations between the secondary and primary sides of the transformer for location of the circuit breaker.

2.8 The Transmission Line

The transmission line may be considered the part of the system bridging the gap between one remote area and another remote area. Its sole purpose is to span long distances. This is accomplished by a system of towers carrying high-voltage conductors. The towers are usually steel and are a familiar sight across the country. The high-voltage systems are as high as 450 kv (450,000 volts). We then get into what is known as the EHV system (extra high voltage). This is in the range of around 700 kv (700,000 volts). The transmission line may be sectionalized for convenience purposes. Also, there may be interlinkages with other transmission systems requiring switching connections. This does

---

[1] See Chap. 6.

not make it a *distribution system*. It is still a transmission system, with the primary function of spanning long distances.

2.9 The Stepdown Transformer

When the transmission line reaches its approximate area of distribution, the voltage must be reduced to a lower level to permit distribution within populated areas. Here, then, we reverse the procedure. The transmission line terminates in a stepdown transformer. The specifications for the transformer are similar to the requirements for the transformer at the generating end. The circuit-breaker requirements also apply; i.e., we may locate a circuit breaker on either the primary or the secondary side.

2.10 The Distribution System

After lowering the voltage at the consumer's end of the transmission system, the power must still be routed to various places of utilization. The *distribution system* is designed to cover a specific area such as a town rather than long distances. The method of distribution is usually by the use of wooden poles. The height of these varies but is usually between 30 and 50 ft. Using poles of this height requires reduced spacing between conductors. Therefore, the voltage is reduced to levels limiting the spacing between conductors to 2 or 3 ft. This allows crossarms of 6 to 10 ft to be utilized. Subtransmission lines are used to connect the transmission system to the distribution system. The voltage on these lines is approximately between 138 and 33 kv. From these main subtransmission lines, a further system of feeders is routed to other specific areas. Again, shorter distances are covered, and therefore the voltage may be reduced. This part of the distribution system is known as the *primary circuits*. From these, we connect further lines known as subfeeders and laterals. The voltages on the feeder part of the system are usually between 13 kv and 4.16 kv. The subfeeders may be three-phase or single-phase, depending on the exact requirements of the load density in that particular area. The single-phase voltages would be the three-phase voltage divided by 1.73 or $\sqrt{3}$. And so the electrical system approximates the shape of a tree, with the transmission line as the trunk and the distribution system as the branches, gradually reducing in voltage.

2.11 The Distribution Transformer

The last link between the distribution feeder and the consumer is the distribution transformer. This transformer is a standard stock item, designed with standard voltages and output ratings. Minor extras are available in the form of *taps* to allow a small percent of adjustment in voltage. Optional impedances are available to permit paralleling with other transformers, and also to limit short-circuit availability to some extent. The transformers are available in single-phase and three-phase; however, the single-phase transformer is more adaptable to general use and provides more flexibility. To obtain three-phase power, it is only necessary to use three single-phase transformers. This also provides a safety margin, in that the failure of one transformer will still allow operation by connecting the remaining two in delta or open wye, providing a service with 58% of the original three-transformer bank output. The voltages, of course, are reduced by the distribution transformer to the standard consumer levels, which could be 115, 230, 277, 480, or 550 volts, depending on local requirements.

2.12 Protection and Control

To prevent failure of power supplies and also to provide protection against excessive overloads, numerous types of devices are used. Basically, all these devices monitor the line variables in one form or other. They eventually result in the operation of a circuit breaker or instrumentation and metering. Some devices operate on current alone, others measure the angle between the current and voltage, and others measure voltage or voltage and current giving apparent power. The different types are too numerous to mention here; however, as we introduce various design procedures, we will also introduce the necessary metering and protective devices required.

2.13 The Substation

The *substation* is the part of an electic system containing one or more power transformers, switchgear, bus-bar distribution, regulating equipment, and (sometimes) a control building. These substations can be a large complex of buildings, steel structures for bus-bar distribution, huge transformers and banks of power-switching circuit breakers, **or** they may be smaller, individual packaged unit substations supplying a small factory. The designation of substation, then, is dependent on function rather than size.

The options available in electric system design are innumerable. Transmission line design, substation design, distribution system design, equipment application, system stability are all individual fields of study, each with its associated pitfalls, shortcuts, dos and don'ts that only experience will give. This does not imply that design cannot be done in all these fields by the competent designer. All the information is available in design procedures, and usually the specifications are outlined either by the customer or by the utility companies in the area. This, then, leaves only the conventional calculations and selection of equipment.

Thus when we consider design procedures, we will assume that the standard practice has already been formulated by the experts and chosen as the normal procedure for all probable conditions. We will encounter unusual conditions where possibly another alternative might be better. However, for reasons not obvious to the nonexpert, it is better to stay with the conventional approach. Again, we do not recommend that original thinking not be used. But it should be pointed out that before deviating from the conventional approach, one should evaluate very carefully the contemplated change and consider all the aspects, from problems of obtaining equipment which may be nonstandard to consideration of the electrician who may have to troubleshoot during a thunderstorm. Sometimes the cost and convenience advantages acquired in straying from the standard are lost many times over in the down time or equipment-replacement costs.

Chapter Three

Job Preparation

3.1 The Single Line

We have considered the basic electric system from the source to the consumer. This is usually the responsibility of the utility company. The common type of design problem encountered by the electrical designer is the distribution of power at the consumer end of the system. The simplicity or complexity of design at the consumer end can go from one extreme to the other. A large steel mill or pulp and paper mill may require its own power system, consuming enough power to supply a small town. Alternatively, the consumer could be a farmer with no greater requirements than those of a farmhouse and barns. But all systems have one thing in common. The basic system is represented by a single drawing, showing the major equipment involved, the interconnections between the equipment, the rating of the equipment, the loads involved, and possibly the metering and instrumentation outline.

This drawing is known as the *single line*. The name is derived from the practice of showing only one line, whereas in actuality three lines are required for a three-phase system. To construct a single line, the designer generally follows a basic distribution practice. The utility company will usually provide a high-voltage source; next, the obvious piece of equipment to install is a transformer. This then requires circuit breakers, interconnecting feeders, subfeeders, power transformers, bus distribution (possibly), and, last, branch circuits connecting to the motors and other equipment. The drawing can then be formed, showing the proposed outline of the design. Actual values of loads or equipment sizes are not absolutely necessary for a preliminary study. In fact, all too often insufficient information is available to make accurate evaluation of equipment sizes possible. This does not affect the basic approach in determining the type of system most suitable for the particular consumer. Alternatives such as radial distribution, ring bus, network or grid system, and parallel or loop-circuit layout may be considered. Automatic throw-over may be a requirement along with the possibility of emergency generators or alternative supply systems.

All these choices can be made without knowing the actual loads and sizes of equipment. With experience, the designer can sometimes make a calculated guess and decide that the main transformer will be of a specific size. This, in turn, will allow him to approximate the quantity and size of feeders to different areas. A list of all motors and other equipment requiring electric power is obtained either from the specifications or from the mechanical design department. Approximate horsepower requirements and approximate power requirements, such as heating and air-conditioning loads, are also supplied. From this information, the experienced designer can evaluate the overall design trend of the project and foresee where a problem may arise.

3.2 The Hold

When a single-line drawing is finalized and ready to issue, before its signing and release, all information on the single line, such as cable sizes, circuit breakers, motor starters, transformers, voltages, transformer connections, metering, instrument transformers, relays and interlocks, etc., should be justified with calculations or instructions from the necessary people in authority.

If the data on the single line is not backed up by the necessary information, then the individual part in question should be circled and noted "Hold." The drawing may then be issued. Anyone having reason to use it will be aware that some of the information is missing, approximate, or tentative. We can show this in the example given in Fig. 3.1 of a small typical single-line diagram in the early stages.

We see from the single line that certain information is given specifically. Also, we see that other information is withheld. How do we arrive at the decision to use a specific size circuit breaker or other apparatus? The first point to consider is what the minimum legal requirements are. Second, we decide whether the minimum requirements are acceptable for this particular situation. The procedure for calculat-

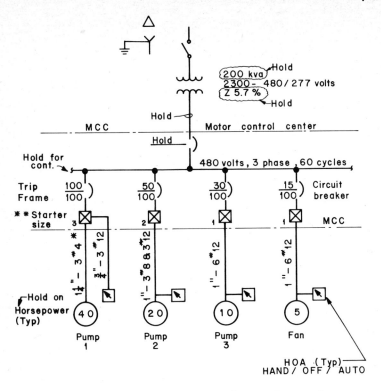

* An optional approach is to run $1\frac{1}{2}"$ conduit with 3 #4 and 3 #12, but refer to standards

** Nonstandard symbol— see standards

Fig. 3.1 Single line (early stages).

ing the minimum requirements is outlined in the National Electrical Code (or local code) for nearly all general requirements. In conforming to our rule of justifying all information on the single line with calculations, we must also know the approximate location of equipment, so that we can determine whether voltage drop[1] is a serious consideration. Once again, it is impossible to know all this information in the early stages of design. However, one thing is available, and that is the site plan or plot plan.

3.3 The Plot Plan

This drawing will show the boundaries of the land owned or available to the client for construction. With this information, we can determine the approximate location of the utility-company power supply. The

---

[1] *IR* for direct current and *IZ* for alternating current. See Sec. 4.3.

obvious location for the main transformer is as close as is practical to this location. Two or three alternate locations can be chosen. When the client conceived the project, he probably had a rough idea of the arrangement of the major buildings and production areas. This is usually shown on the plot plan. With this information available, the designer can segregate the areas by density of equipment. Without actually knowing the power requirements of an area, he can at least anticipate the approximate routing of feeders and approximate locations of power-distribution centers. When this is established, approximate distances can be determined.

Considering the worst conditions and longest distances, voltage-drop calculations can be carried out. The distances are approximate, it is true, but the question is what percentage of error are we dealing with. First we consider the longest distance, and then we evaluate the voltage drop. We know from our study of logic (see Chap. 19) that an assumption may be right or wrong—it does not matter—the conclusion is the important point. With this in mind, we find we have only two possible answers: either we have a problem, or we do not. Because our first consideration is the worst possible condition, we may find that we do not have a problem. In this case, we just file our sample calculation and forget it. The other alternative, however, is that we do have a problem. What now? The problem now becomes a matter of economics. Are we considering conductors of large size, requiring special order, or are we considering small wire, which is available at very low cost? In the latter case, we do not waste valuable designer's time (this also costs money). We evaluate the next standard size and, if satisfactory, we specify it on the single line and file the calculations. The first consideration, however, requires more specific information. In this case, the size is evaluated, based on the worst condition being valid. The information is then entered on the single line with a "Hold," indicating the possibility of a change.

This procedure can be followed for the sizing and selection of nearly all the equipment required on the single line. The choice of sizes is limited by the range of standard equipment available. Usually, the range between the one selected and the next standard size determines the selection. With the single line and plot plan roughed out in the manner we have just described, we have a broad picture of the scope or size of the project. From this, the experienced designer should have an instinctive feeling as to whether to classify the project as small, medium, or large. Also, he will know the time limits for design, such as "No rush," "Steady work," or "Rush"—and, of course, the ultimate requirement of "We want it yesterday." With this information, the next approach is to formulate the design "steps."

3.4 The Drawing List

With the aid of the previously outlined information, the designer must make a mental picture of all the drawings required to complete the project. The basic requirements are standard. The question in mind is, How many? The plot plan serves as a guide. The standard scale for electrical drawing is $\frac{1}{8}$ in. = 1.0 ft or $\frac{1}{4}$ in. = 1.0 ft, and the standard size paper for a particular project is usually known. Therefore, the number of drawings to cover an area can be determined approximately. In refinery or chemical plants, for example, the system of conduit banks, either underground or above-grade, can usually be routed very accurately. A certain amount of details and sections will be required for the conduit banks. Therefore, a number of drawings should be allowed to cover this work. A particular high-density area of equipment should be contained on the one drawing, if possible; if not, then more drawings may be necessary.

If the project is very large, the plot plan should be covered with grid lines representing drawings. Various levels of platforms and floors —all must be considered and allowed for in the correct sequence. Sometimes the numbers for the drawings are supplied in the form of a block, for example, E100 to E300. In this case, each drawing is given a number between E100 and E300. Another possibility is the use of a particular drawing number for a certain portion of the project, with subnumbers for each drawing. An example of this would be a requirement of four drawings for underground conduits. The first drawing would be E100-1, the second would be E100-2, etc. This allows a certain flexibility, in that an extra drawing can be added, if required, without upsetting the sequence of numbering. For instance, if the numbers allowed are E100 to E200, and we decide that four drawings are necessary for underground distribution, we assign E100 to E104. We then assign E105 to E110 for above-grade work. Later on, in the progress of design, we find that, through some new requirement, we need another underground drawing. What number do we use? We should use E105. This continues the sequence of underground distribution, and now we must alter all the numbers on the other drawings, changing E105 to E106, and so on.

This is inconvenient but not a serious change, except for the fact that throughout a design, references are made by drawing number to details, sections, and various other items. Now we do have a very serious problem: Do we change all the numbers and put the new drawing in sequence, or do we leave the drawing numbers as they are and put the new drawing out of sequence? This is a decision requiring an evaluation of the overall advantages and disadvantages. The obvious advice here is, *do not* get into this trap. Leave considerable spare numbers between sections of the project drawing numbers. Identify them on the drawing list as spare, or assign them to a particular section. This precludes the assigning of them to some other section of the project. When working on specific projects, a checklist should be made for all basic sections of design encountered in the particular project. An example of a drawing checklist would be as follows:

DRAWING CHECKLIST—TANK FARM

1. PLOT PLAN
2. SINGLE LINE
3. SUBSTATION
4. UNDERGROUND CONDUITS AND DETAILS
5. ABOVE-GRADE CONDUITS AND DETAILS
6. EQUIPMENT LAYOUT AND DETAILS
7. AREA LIGHTING AND DETAILS
8. LOADING RACKS AND DETAILS
9. OFFICES AND DETAILS
10. CONTROL AND INSTRUMENTATION
11. GROUNDING
12. CORROSION SYSTEM
13. PUBLIC ADDRESS AND TELEPHONE SYSTEM
14. FIRE ALARM SYSTEM
15. EMERGENCY SYSTEMS
16. (CONTINUE WITH OTHER ITEMS)

And so the list grows. If an item is not applicable, it is easy to check it off as "not applicable." However, it is much harder to find, halfway through a design, that someone neglected to mention that a corrosion prevention system would be required. With the checklist (which grows as one gains experience), all these things are taken into account; if a question arises, a written request is made for an answer. The answer is then retained in the job file.

3.5 The Job File

It should be obvious by now that a complete job file is necessary. Never discard information, no matter how insignificant or inapplicable it may seem. The familiar saying, "Write it—don't say it," is a good motto to follow when considering a job file. The first step in any design procedure is to set up a job file. This is obviously created so that someone other than the usual designer may have all the information available to continue the design in the absence of the usual designer. It also serves a secondary purpose of substantiating de facto instructions at some future date. We must accept the fact that although ethics are professed, they are sometimes accidentally "bent" a little. We must also accept the fact that "passing the buck" is not new. It is used in politics, and in every form of business. Many times it is unintentional; however, the fact remains it is there and we must live with it. Therefore, it is essential that all information pertaining to the project be retained where practical. We cannot see into the future. Also, we are all fallible. Therefore, mistakes are to be expected. The problem is to know how to handle them when they do occur, without detriment to the employer, the client, and the project. The job file is invaluable at a time like this.

Many times, a two-minute conversation with the correct person eliminates any further problem. Without the job file, finding this person may take considerable time; also, false information may be accumulated along the way, at the expense of the employer, the client, and the job.

Sometimes the designer is placed in a quandary with regard to his responsibility and action. Should he do exactly as instructed, or should he point out a possible "flaw" in the instructions? This is a problem. Certainly an obvious error should be pointed out, but at the same time, alternative approaches are innumerable and a source of irritation. When confronted with this type of decision, one should ask oneself the question, Is it really significant? When this is established, usually the question of action and responsibility is also established.

Chapter Four

Motors and Exciter Wiring Calculations

4.1 Basic Requirements [NEC 430-1(a)]

The basic requirements for a motor circuit are

1. MOTOR-BRANCH-CIRCUIT OVERCURRENT PROTECTION (NEC 430-51)
   a. Circuit breakers (preferred)
   b. Fuses (use with high fault currents)
2. MOTOR CONTROLLER (NEC 430-82)
   a. Magnetic starter—full-voltage type
   b. Magnetic starter—reduced starting current—various types
   c. Magnetic starter—reversing type
   d. Magnetic starter—two-speed type
   e. Manual starter—various types
3. COMBINATION STARTER (see manufacturer's catalog)
   a. Items 1 and 2 in a single-unit package
4. MOTOR-RUNNING OVERCURRENT PROTECTION (NEC 430-32)
   a. Separate overcurrent device
   b. Thermal protector integral with motor
   c. Embedded temperature detectors
   d. Combinations of (a), (b), and (c)
5. MOTOR-BRANCH-CIRCUIT CONDUCTORS
   Recommended *minimum* No. 12 AWG (TWH or RHW)—75°C maximum operating temperature

6. MOTOR-DISCONNECT DEVICE (NEC 430-86)
   a. Motor-circuit switch—rated in horsepower
   b. Circuit breaker (with or without auto trip)
   c. General-use switch—limited conditions
      (NEC 430-109, Exceptions 2 to 4)
   d. Isolation switch—limited conditions
      (NEC 430-109, Exception 4)
   e. Service switch—limited conditions
      (NEC 430-106)
   f. Lockout on motor controller [NEC 430-86(a)]
   NOTE: *The normal "plug in" motor-control-center combination starter meets NEC 430-86(a) and is the normal method of installation in industrial plants.*
7. MOTOR-FEEDER CONDUCTORS (grouped motors)
   a. See item 5 and NEC 430-24
   NOTE: *A motor feeder does not usually feed a motor. Its function is to feed a central location, such as a motor control center, which in turn feeds individual motors with branch circuits.*
8. MOTOR-FEEDER SHORT-CIRCUIT PROTECTION (grouped motors)
   a. Circuit breaker
   b. Fuses [NEC 430-62(a), (b)]

4.2 Squirrel Cage (Three-phase Motor)

The standard "squirrel-cage" motor requires only three conductors to connect the motor to the power source. The control circuit will usually require three conductors, depending on the control at the motor. For the purposes of this book, and also in line with routine practice, we will only consider an installation with a HAND/OFF/AUTO[1] control at the motor. The section on controls will cover alternative approaches.

4.3 Three-phase Induction-motor Calculations

As a typical example, we can have a 25-hp pump located in a pump house and remote from the motor control center supplying the power. We can assume the distance between pump and motor control center to be 500 ft. The following method should be a guide to calculation procedures generally. It is the same approach we used in our mathematical calculations:

1. 25 hp at 440 volts—three-phase   Original specification
2. Full load amperes = 34   NEC Table 430-150 (mandatory)

[1] A STOP/START push-button station also requires three conductors.

3. Branch-circuit conductor =   NEC 430-22 (mandatory)
   $34 \times 1.25 = 43$ amp
4. Wire size, 43 amp: No. 8 AWG THW   NEC Table 310-12
   at 75°C
5. Conduit, rigid steel: use three No. 8   NEC Tables 1 and 3A
   AWG ¾-in. (Hold)   (mandatory)
6. Underground: use minimum 1-in.   Trade practice
   rigid steel (Hold)
7. Magnetic starter: NEMA size 2   Manufacturer's catalog
   FVNR (combination type)[1]
8. Branch circuit breaker 70-amp trip   NEC Table 430-153
   amperes: $34 \times 250\% = 85$ amp
   (maximum)
8a. Branch circuit breaker frame: See manufacturer's catalog and
    100 amp   short-circuit rating
9. Control: three No. 12 AWG or,   Trade practice
   No. 14 AWG THW
   HAND/OFF/AUTO (optional)
   or STOP/START
10. Conduit will contain three No. 8   Trade practice
    AWG and three No. 12 AWG; combine lines 4 and 9
10a. Conduit: 1-in. for surface runs   Trade practice[2]
10b. Conduit: 1-in. for underground runs   Trade practice (normal)
    (minimum)
11. Disconnect required at the motor   NEC 430-86(a)

or

11a. Lockout on combination starter   NEC 430-86(a)
12. Motor-running protection (heaters in   NEC 430-6(a), NEC 430-32(a1)
    controller); nameplate amperes required but not available. Use line 2,
    i.e., 34 amp, then $34 \times 115\% = 39$ amp
13. Check voltage drop, that is, 3% for   NEC 210-6(d)
    power loads
13a. IR drop = voltage drop = L = length in feet, R = resistance[3] per foot, 1.73 for three-
    $34 \times L \times R \times 1.73 =$ phase, 2.0 for one-phase
    $34 \times 500 \times \dfrac{0.6}{1,000} \times 1.73 =$
    16.6 volts
13b. $(16.6 \times 100)/440 = 3.78\%$ at normal power factor[4]

[1] FVNR—Full-voltage nonreversing with circuit breaker and starter in one unit.
[2] See Table A.6.
[3] This actually should be impedance; that is, $Z = \sqrt{R^2 + X^2}$, but R is close enough (see NEC Table 8).
[4] PF—see Sec. 7.12.

13c. As the induction motor would probably be lower than normal PF and we are over the 3%, we should go to the next size conductor, which is No. 6 AWG

13d. New conduit size: three No. 6 AWG and three No. 12 AWG—use 1¼-in. rigid steel

14. Check new wire size: IR drop = $34 \times 500 \times 0.4 \times 10^{-3} \times 1.73 = 11.8$ volts

14a. $(11.8 \times 100)/440 = 2.7\%$. This at normal PF.

The previous example is shown in single-line representation in Fig. 4.1.

NOTE: *The break point between power and control in single and individual conduits depends on the relative wire size. Usually any conductors below No. 2 AWG can use a single conduit. Number 2 conductors and above will require separate conduits, because the smaller conductors could be damaged when pulled in with heavier ones.*

4.4 Multiple Motors

Now we come to the situation where we have more than one motor in a group. Here we will have a few more calculations to complete, but nothing more difficult than the calculations we have just performed. Consider a pump house containing four motors and two future motors. The sizes will be 10, 10, 15, and 25 hp. The two future motors will

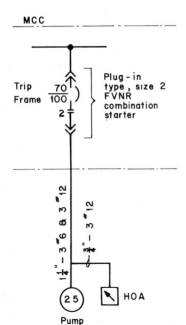

Fig. 4.1 Single-line representation.

be 25 and 40 hp. How do we approach this installation? First, we draw a single-line diagram with all the main components. We know that the future motors will definitely be installed; therefore, we treat these as though we were actually installing them. We will consider an underground installation of conduits. We will also install a motor control center in the pump-house building, with a feeder from the main plant, supplying the motor control center with 3-phase 460-volt power. The multiple motors are shown in single-line representation in Fig. 4.2.

The calculations for the individual motor branch circuits are simply repetitions of our preceding example; and with underground installation specified, we will use a minimum conduit size of 1 in. The breaker frame is selected as 100 amp. At this point, the reader should accept this value. When we get into short-circuit studies, we will explain the selection of the breaker frame size. The breaker trips are based on the standard breaker having approximately 200% of full load current. This is to prevent tripping on start-up; it also provides some sensitivity rather than the maximum allowed being installed. This also reduces

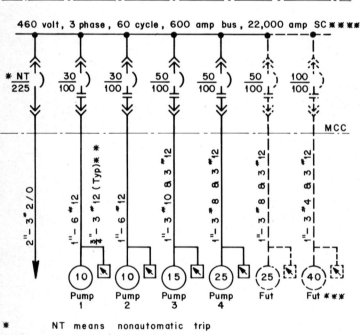

* NT means nonautomatic trip
** Typ means typical
*** Fut means future
**** SC means short circuit

Fig. 4.2 Single-line pump house.

the cost somewhat, because the cost of a circuit breaker increases with current rating. Remember that the circuit breaker is for the protection of the circuit—and *not* for protection of the motor. The breaker is only for protection against overcurrents due to short circuits and ground faults.

In our single-line diagram, we have shown a circuit breaker as a main disconnect. However, this disconnect does not have to be a circuit breaker. In fact, it does not have to be there at all. Consider the situation: We have a motor control center in a pumphouse. Shutting down all the equipment in case of emergency requires six separate operations when all motors are installed. It is, therefore, good practice to be able to shut down the whole pump house with one fast operation if necessary. It is usually worth the added expense when trouble is encountered under routine operation. The difference in time advantage may mean the saving or losing of a bearing, pump, motor, or human limbs. This circuit breaker is in "cascade" with the main feeder breaker. We ignore this fact and we size the breaker with a trip as large as or larger than the main feeder breaker, depending on our aims. We will size them both so that if a short circuit occurs in the pump house, the breaker in the main power supply will trip. This will serve the purpose of warning the operator in the main area that something is wrong in the pump house. With this in mind, we will calculate the main breaker first, and then use a nonautomatic trip for the main disconnect. The requirement for a feeder-breaker size is that we take the size of the breaker for the largest motor and add the full load currents of the other motors. This, then, gives us the maximum size allowed.

4.5 Future Motor Installations

Now we encounter a small problem: Do we consider the future motors or not? The second question is based on the first: Are we going to install a feeder large enough to take the future load? To evaluate this problem, we will bear in mind that it is an underground installation, and that once the raceway is installed, we cannot change it without considerable expense. Therefore, always try to install for the future if the client will allow the extra expense. There is always a problem between accounting and engineering; and sometimes the decision is not an engineering decision, but is dictated by client accounting. If the answers to our questions are not readily available, make the decision. We will decide to install for the future. Therefore, our feeder size will be large enough to carry the two future motors, and this will also have a bearing on the selection of the breaker trip. Following routine calculation procedure, we have

BREAKER TRIP—MAIN FEEDER TO PUMP HOUSE

1. Motors: 10, 10, 15, 25, 25, and 40 hp — Single-line diagram
2. Full load amperes: 14, 14, 21, 34, 34, and 52 — NEC Table 430-150
3. Largest motor: 40 hp at 52 amp — NEC 430-62(a)
4. Largest motor trip (circuit breaker): $52 \times 250\% = 130$ amp — NEC Table 430-152
5. Feeder-breaker trip: $130 + 34 + 34 + 21 + 14 + 14 = 247$ amp — NEC 430-62(a)
5a. Next standard size: 250/400-amp frame (maximum) — NEC 430-62(b)

This, then, is our maximum allowable trip for feeder protection. Use 225-amp frame and 200-amp trip. Now we must consider the conductor size for the feeder. Here, we take 125% of the largest motor and add the full load currents of the other motors.

CONDUCTOR SIZE—MAIN FEEDER TO PUMP HOUSE

1. Motors: 10, 10, 15, 25, 25, and 40 hp — Single-line diagram
2. Full load amperes: 14, 14, 21, 34, 34, and 52 — NEC Table 430-150
3. Largest motor: 40 hp at 52 amp — NEC Tables 430-150 and 430-24
4. $1.25 \times 52 = 65$ amp[1] — NEC 430-24
5. $65 + 34 + 34 + 21 + 14 + 14 = 182$ amp — NEC 430-24
6. 182 amp; use three No. 3/0 AWG 75°C THW conductors — NEC Table 310-12
7. Conduit size: 2-in. rigid steel — NEC Table 3A

4.6 Breaker Trip Setting

Now, to determine which trip setting we should have on the feeder, we consider the preceding calculations. In line 5 of main-feeder-breaker calculations, we have 247 amp. This is based on all motors (future included). In line 5 of conductor size calculations, we have 182 amp, also based on all the motors being installed with a standard trip of the "next highest rating" (NEC 240-5, Exception 1), which would be 250 amp allowable. However, we already selected a 200-amp trip. For the motor control disconnect, we can use a circuit breaker without auto trip.

In NEC 430-62(b), the statement is made that *"where heavy capacity feeders are installed to provide for future additions or changes, the feeder overcurrent protection may be based on the rated ampacity of the feeder conductors."* The connective word "may" is not a mandatory instruction (NEC 110-3); therefore, we can size the breaker at 200 amp if we choose.

[1] See NEC Example 8.

4.7 Code Compliance

Code violation? Consider NEC 430-62(a) for motor-feeder short circuit and ground fault protection. The Code says, "A feeder which supplies a specific fixed motor load . . . ." In our case, the actual specific motor load (installed[1]) would be 10, 10, 15, and 25 hp. And, in this case, the breaker trip would be

1. Motors: 10, 10, 15, and 25 hp — Original
2. Full load amperes: 14, 14, 21, and 34 — NEC Table 430-150
3. Largest motors: 25 hp at 34 amp — NEC Table 430-150
4. Largest motor trip (circuit breaker): 34 × 250% = 85 amp — NEC 430-63(a)
5. Feeder-breaker trip: 85 + 21 + 14 + 14 = 134 amp — NEC 430-62(a)
6. Use 150-amp trip/225-amp frame — NEC 240-5, Exception 1
7. Conductors installed: three 3/0 THW at 75°C (200 amp) — See feeder-conductor calculations
8. 200-amp circuit-breaker option — NEC 430-62(b)

4.8 A Logic Analysis

Now we come to the question, Would we have violated the Code by specifying the 250-amp trip? By the wording of the Code, we are protecting the conductors against a short-circuit condition—not an overload on the motor (this is protected by another device). The size of the conductors is based on the size of the motors, and the size of the breaker is in relation to the size of the motors. Then by our logic method of analysis, we should be able to determine whether there is a Code violation or not. First, we must list the requirements for sizing the circuit-breaker trip for the feeder. We do this as follows:

Proxy (a). A feeder supplying a specific motor load
Proxy (b). AND consisting of conductor sizes based on NEC 430-24
Proxy (c). SHALL be provided with overcurrent protection NOT greater than the largest rating OR setting of the branch-circuit protective device for any motor of the group (based on NEC Tables 430-150 and 430-152) AND plus the sum of the full load currents of the other motors of the group
Proxy (d). Correct feeder-circuit-breaker trip rating

To set up our logic equation, we can represent the parameters by using the identifying letter or "proxy." We now ask ourselves *if we have*

---

[1] The Code does not specify "installed." It states "specific fixed motor load," and it does not state a specific time, i.e., now, later, or any other time.

the correct trip size; then we must comply with the previously outlined conditions. We recognize the word "then" to be an implication. Thus our expression will be in the form of an implication.

1. If d THEN $a \cdot b \cdot c' = T$ Hypothetical proof

We assume that we have complied with all conditions in the proposition; therefore, we use T.

2. $d \rightarrow a \cdot b \cdot c' = T$ Substitute operation sign

We have selected a trip we think is correct; therefore, we replace d with T.

3. $T \rightarrow a \cdot b \cdot c' = T$ Substitute in line 2

From our laws of implication, if the proposition is true and the premise is true, then the conclusion must be true for the expression to be valid. So, we consider a. Do we have a feeder supplying a *specific* motor load T or F? This condition is met because we set the motor load at 10, 10, 15, 25, 25, and 40 hp, no more and no less. This would comply with "specific." No mention is made of time, and it is a fact that we are supplying a specific load of motors. Therefore, a is T.

4. $T \rightarrow T \cdot b \cdot c' = T$ Substitute in line 3

Now, we consider b. We can show by our previous calculations that this is T.

5. $T \rightarrow T \cdot T \cdot c' = T$ Substitute in line 4
6. $T \rightarrow c' = T$ Law of redundancy

Now we are left with the last consideration. Notice that we express the parameter c as negated. This is a result of the NOT, i.e., "overcurrent protection NOT greater." However, in logical equations, it is expressed in the affirmative and then negated. We also see that c is a compound expression, containing a logical sum and product. We can consider this mentally and realize that we must meet the first or second condition, and also the third condition. The question is, Did we? We are told to use NEC Tables 430-150 and 430-152, and so now we have 40 hp at 52 amp $\times$ 250% = 130 amp, plus full-load currents of the others, which would give 130 + 117 = 247 amp. Now the statement says "NOT greater than 247 amp." Under NEC 240-5, Exception 1, we can go to the next size higher if it is a nonstandard rating. Therefore, we can go to 250 amp. Thus, in the last line of our logical analysis, we replace the c' with a T, since we have complied with all the conditions.

7. $T \rightarrow T = T$ Substitute in line 6

We have justified our original hypothesis in line 1 by arriving at a valid expression in line 7. If the a, b, or c' could not have been given a valence of T, then we would have had, in line 7, $T \to F = T$, which, from our laws of implication, is not valid. It would also state that our original hypothesis in line 1 is incorrect. Therefore, it seems that we do not have a code violation.

This is a good example of how practical use is made of mathematical logic in everyday problems. One thought to bear in mind is that the local inspection department makes its own interpretation of the code. Therefore, when the code is in question, obtain a ruling in writing from the local inspection department before proceeding with a controversial design. An analysis like the one we have just completed would be good for an appeal in an "after-the-fact" adverse ruling, involving expensive changes. If a prior ruling is given, even if it does not seem correct, it is usually better to accept it, if it does not involve any major added expense.

4.9 The Wound-rotor Motor

Up to now, we have considered only the conventional squirrel-cage motor. As previously explained, the "squirrel cage" identifies the type of rotor in the motor; "wound," of course, designates another type of rotor. To digress from our subject briefly, the squirrel-cage rotor is

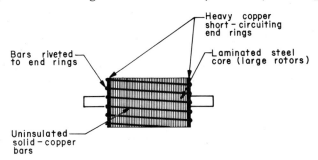

Fig. 4.3 Squirrel-cage rotor.

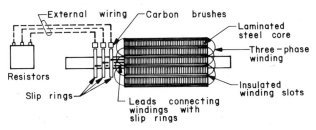

Fig. 4.4 Wound rotor.

essentially a solid rotor with embedded bars on the periphery which are joined at the ends by a solid ring. The remaining bulk (iron) of the rotor can be of cast or laminated construction. There are no external connections to the squirrel-cage rotor. In contrast, the wound rotor (three-phase) consists of a three-phase double-layer winding, a modified wave winding connected in star or delta and chorded or fully pitched.[1] The ends of the three-phase winding are connected to slip rings, and conductors are then connected to the brushes on the slip rings and to a bank of resistors. Resistance of the rotor can then be varied as required (this will be discussed later).

4.10 Secondary Conductors

Our only concern at this point is how to size the conductors between the resistor bank controller and the slip rings on the rotor. The code requirements (NEC 430-23) are given for "continuous duty" and "other than continuous duty." When one is considering a new installation, it is suggested that only the "continuous-duty" rating be used. The "other-than-continuous-duty" rating becomes useful when one is "soft starting" or considering increasing an existing installation. It allows a higher rating for the existing installation when considering time-rated motors.

The Code requirement for continuous-duty rating of the conductors, between the slip rings and the resistance controller, is that they be "125% of the full load current of the secondary (rotor) winding." This is a very straightforward statement. However, in the early stages of a project, the only information available is that a wound-rotor motor is required with an approximate horsepower specified. The motor may not have been purchased yet; therefore, the manufacturer is unknown. This, then, leads to the problem of how to obtain the secondary current to size the conductors. In a situation like this, the conduit and/or conductors are shown on the drawing but are designated with "Hold," as shown in Fig. 4.5.

[1] Refer to a motor-design handbook.

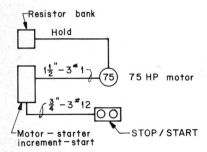

Fig. 4.5 Wound-rotor motor wiring 460-volt three-phase increment start.

If the designer is experienced in motor design, he can possibly try to evaluate the secondary current (this is not recommended). However, we will give the necessary equations for calculating the secondary current and rheostat resistance:

Secondary current $$I_r = \frac{\text{hp} \times 746 \times k_2}{E_r(1 - \text{slip})3}$$

Slip-ring voltage[1] $$E_r = \frac{\text{hp} \times 746 \times k_2}{I_r(1 - \text{slip})3}$$

Rheostat resistance $$R_{\text{rh}} = \frac{E_r(1 - \text{slip})}{I_r k_3}$$

where E_r = rotor voltage at standstill between slip rings
I_r = full load rotor current per phase
R_{rh} = rheostat resistance per phase for full load starting torque
k_2 = 1.73 for star connection
k_2 = 1.00 for delta connection
k_3 = 1.73 for star connection
k_3 = 3.00 for delta connection
Slip is expressed as a decimal, that is, 0.03 for 3% slip

The evaluation of E_r is dependent on approximately eight design variables. Therefore, it is highly improbable that a calculated guess would result in the same current values as those on the nameplate of the actual motor.

4.11 Conductor Sizing

If it is absolutely necessary to specify a conductor size, then the motor manufacturer should be contacted and an approximate value for E_r can be obtained, or even an approximate I_r, in which case a conductor size can be calculated using a 3% slip. Multiplying by 125% will give the Code-required conductor size. As a safety margin, and based on the fact that the information is only approximate, install a conductor of the next higher size and mark "Hold." Where the secondary resistor bank is separate from the controller, then the ampacity of the conductors between controller and resistor shall be not less than that given in NEC Table 430-23(c). The section of the code for wound-rotor secondaries is NEC Tables 430-23, 430-32(d), and 430-82(c). In actual practice, a rheostat-type resistor control is not necessarily used. A step-type control, which will allow starting with all resistance "in," can be used. After

[1] Note that the motor may be a high-voltage motor, for example, 2,300 volts, but the secondary will be of a lower voltage.

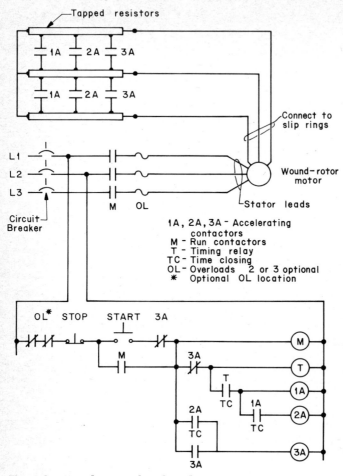

Fig. 4.6 Wound rotor—three-line diagram.

a short period, half the resistance will be cut out. Then, as the motor reaches full speed, all the resistance will be cut out, leaving a short-circuited secondary, which will have most of the characteristics of the conventional squirrel-cage motor. A diagram of a typical circuit for the above type starter is shown in Fig. 4.6.

4.12 The Single-phase Motor

The single-phase motor is usually limited in size to a maximum of 2 hp. Above this, three-phase motors are used.[1] There are various types of motors, such as split phase, capacitor start, capacitor motor, repulsion

---

[1] In actual fact, process plants may use three-phase motors at ½ hp and up.

motor, shaded pole, and series motors. The majority of these are in the "fractional-horsepower" range and are used in the ordinary appliances and tools. These include washing machines, electric drills, blenders, electric clocks, and innumerable other items. In many cases, the controls are furnished with the appliance. Therefore, all that remains is to provide the branch-circuit conductors. The fact that the motor is single-phase has no bearing on the conductor rating. The procedure is the same as for the three-phase motor. We obtain the full load current from NEC Table 430-148, and then we comply with the applicable sections of NEC 430-22 and/or NEC 430-24, and NEC 430-37 for the overcurrent requirements.

The controller for a single-phase motor may be automatic, i.e., magnetic starter and remote push-button control, or it may be manual. In the latter case, a device similar to a switch is used. It contains an overcurrent mechanism in addition to the switching contacts. In certain cases if a motor has a plug connection, the plug may serve as a disconnect device. The trend toward overcurrent protection integral with the motor in the fractional-size motors is becoming more evident. Therefore, this must also be considered when conforming with NEC 430-32 (overload protection). In the small-size motors, more than one motor may be connected to a branch circuit. This is covered in NEC 430-53. This is permissible under limited conditions. When the motor is used in a process application rather than in a convenience application, individual motors on individual circuits are recommended, with individual conduits where economically practical.

A typical circuit for a single-phase motor is shown in Fig. 4.7. One important consideration should be made when dealing with single-phase motors: They should be categorized. An important motor, such as a proportioning pump or other process or industrial application, has as much importance as a 400-hp vacuum pump. Thus the motor should be treated as if it were a large motor. A magnetic starter should be used for control, and individual installation is recommended. The starter should be located in the motor control center with the other process

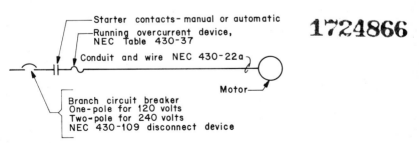

Fig. 4.7 Single-phase motor wiring—single-line diagram.

equipment starters, and the single-phase power should be supplied from a small transformer and panel located in the motor control center.

When the motor is not of great importance, such as an air-circulating fan which is turned on and off at will by personnel in the immediate vicinity, then the power may be obtained from a lighting panel in the area. The controller may be manual and located at some convenient spot in the area of the fan itself. The important point is the evaluation of the motor into one or the other category.

A hazardous area presents a slight problem when one is dealing with single-phase motors. The majority of such motors have a starting switch on the rotor operated by centrifugal action. This constitutes an arcing device, and it requires that the motor be explosionproof. This may make the cost prohibitive. When this situation arises, the normal procedure is to consider a three-phase motor of the same rating or even slightly larger. This may allow the use of an open-type motor in some hazardous areas. Also, the cost should be reduced for purchase of a standard open-type motor, as against an explosionproof single-phase motor, not to mention more satifactory delivery schedules in the case of the former. When using a motor larger than required, the main drawback would be low power factor. However, it is usually more economical to accept this in lieu of going to an explosionproof motor.

4.13 Direct-current (DC) Motors

The dc motor requires two conductors and one overcurrent device, and also branch-circuit protection in one or both conductors, i.e., two-pole as in NEC Tables 430-37, 430-147, and 430-152. The dc motor consists of two windings, one known as the *armature* on the rotating part, and the other known as the *field winding* on the stationary part. The various ways of connecting these windings results in three basic types of dc motors. These are *shunt wound, series wound,* and *compound wound.* There are variations in each type, but essentially there are only three general types. When a dc motor is at rest, the armature is like a very low resistor across the line. As the motor builds up speed, it generates a back EMF,[1] which limits the current. During the initial few seconds of starting, we have to insert some resistance into the armature circuit to limit the starting current. This resistance is then cut out in steps until it is completely removed from the armature circuit. This type of controller must be self-resetting. If, for any reason, the motor is stopped, then the controller must automatically revert to its position of inserting the resistance in the armature circuit. See NEC 430-82(c), which gives the requirements for rheostats and under-voltage release.

[1] Electromotive force.

The speed of a dc motor is nearly directly proportional to the impressed voltage, and inversely proportional to the flux per pole. By controlling the field current, the flux can also be controlled—thereby controlling the speed. Unfortunately, an open circuit in the field will cause a condition of weakest flux and highest speed. This can become dangerous, and it presents a problem in some motors, and so the Code requires certain limitations on variable-speed motors and certain dc motors. This is covered in NEC 430-88 and 430-89. Apart from these limitations, the calculations follow the normal procedure for branch-circuit conductors, protective devices, etc. Using a 240-volt 25-hp shunt motor as an example, we will have

1. 25 hp—240 volts dc — Original
2. Full load amperes: 89 amp — NEC Table 430-147
3. Conductor size: $89 \times 1.25 = 111$ amp— No. 2 AWG THW 75°C — NEC 430-22, Table 310-12
4. Conduit size: two No. 2 AWG—; use 1¼-in. — NEC Table 310-3A, STOP/START
5. Starter: NEMA size 3—dc 3, accelerating points — Manufacturer's catalog
6. Branch circuit breaker 150 amp—two pole: that is, $111 \times 150\% = 167$ amp — NEC Table 430-152
7. Check voltage drop: $89 \times L \times R \times 2 =$ volts — L = length—one-way in feet, R = resistance per foot
8. Control: allow three No. 12 AWG in ¾-in., if unspecified — STOP/START
8a. Motor protection: $89 \times 130\% = 115$ amp maximum — NEC Tables 430-37 and 430-34

We see, then, that the procedures for dc and ac calculations are similar. In the case of a separately excited motor, this will usually be a completely designed system, and the manufacturer will supply the drawings and information on any conductors involved.

4.14 Synchronous Motors

The synchronous motor is a true constant-speed motor. The speed of this ac motor is dependent on the number of poles and the frequency of supply. The equation for determining the speed is

$$\text{rpm} = \frac{f \times 60 \times 2}{p}$$

where f = frequency
p = number of poles
rpm = revolutions per minute

Considering an example of a four-pole motor (60 cycle), the speed is

$$\text{rpm} = \frac{60 \times 60 \times 2}{4} = 1{,}800$$

This is the actual speed of the rotor—there is no *slip* to consider.

The design of the stator winding of the synchronous motor is similar to that of the squirrel-cage motor. Therefore, the conductors supplying the stator require adherence to the same rules. The design of the rotor, however, is entirely different. The motor is not normally self-starting. However, by unique design, a partial squirrel-cage-type winding[1] is built into the rotor. This winding then is used for starting, bringing the motor up to nearly synchronous speed. The speed of a *squirrel-cage motor* is nearly the same as the speed of a synchronous motor—except that the rotor will *slip* with respect to the rotating field of the stator. The equation is

$$\text{rpm} = \frac{f \times 60 \times 2}{p} - \text{slip speed (rpm)}$$

or

$$\text{rpm} = \frac{f \times 120}{p} (1 - s)$$

where s is percent slip, expressed as a decimal. Consider an example of a four-pole (60-cycle) 2% slip.

$$\text{Rotor speed} = \frac{60 \times 120}{4}(1 - 0.02)$$
$$= 1{,}800 \times 0.98$$
$$= 1{,}764 \text{ rpm}$$

We see, then, that the *squirrel-cage* motor is not a *true constant-speed motor*. This effect is used for starting the synchronous motor.

4.15 DC Rotor Field

Now we have a situation where the synchronous-motor rotor is brought up to speed as a squirrel-cage motor. The synchronous-motor rotor requires a supply of dc power. This is provided through two slip rings and brushes mounted on the shaft,[2] connected to two wires, which,

---

[1] Amortisseur winding.
[2] See the manufacturer's catalog for the "brushless" type.

in turn, connect to the dc source of power. When the speed of the rotor is nearly at synchronous speed, then (through automatic switchgear) the rotor is energized with dc power. This will allow it to pull into synchronism. It will then rotate at synchronous speed, according to our equation. The motor will maintain this speed regardless of load (within design limitations).

For a particular motor, and considering a fixed load, there is one adjustment of dc field, which will result in a minimum stator current.[1] Now we run into more questions. As designers, the only information we will be supplied is probably horsepower and synchronous type. From this information, we must proceed with the design. But how do we evaluate the stator current and rotor current? First, we construct a preliminary single-line diagram for our own study purposes. From

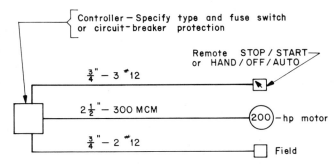

Fig. 4.8 Synchronous motor wiring 460-volt three-phase unity power factor—preliminary sketch.

this we see that three sets of conductors must be run between the motor and the controller. These are control, field, and stator conductors. If we are not sure of the type of control to be used at the motor, then we assume a HAND/OFF/AUTO control with three No. 12 AWG conductors in ¾-in. conduit, and we mark "Hold" on the drawing, until we confirm or change the design. Now, we consider the stator full load current. According to NEC 430-6(a), we use NEC Table 430-150 for determining full load currents. Under "Synchronous Type Unity Power Factor," the full load currents are specified. Considering an example (200-hp 440-volt three-phase 60-cps 100% PF), our calculations will follow in our usual form:

1. 200 hp at 440 volts—three-phase (60 cps at 100% PF)   Original
2. Full load amperes: 210   NEC Table 430-150, synchronous motor

---

[1] See a motor-design handbook on "vee curves."

40 Electrical Design Information

3. Branch-circuit conductor: 210 × 1.25 = 263 amp
   NEC 430-22
4. Wire size: 300 MCM, THW 285 amp at 75°C
   NEC Table 310-12
5. Conduit: rigid steel 2½-in.
   NEC Table 3A
6. Magnetic starter: NEMA size 6 FVNR
   Contact manufacturer
7. Branch-circuit *fuses:* 210 × 300% = 630 amp maximum
   NEC Table 430-152
8. Running protection: 210 × 130% = 273 amp maximum
   NEC Table 430-34, contact manufacturer
9. Control: Use three No. 12 AWG STOP/START or HAND/OFF/AUTO
   Trade practice
10. Conduit-control: use ¾-in. for three No. 12 AWG
    Trade practice
11. Disconnect required at motor
    NEC 430-86(b), contact manufacturer

or

12. Lockout on combination starter
    NEC 430-86(a), contact manufacturer
13. Check voltage drop: 3% for power loads
    NEC 210-6(c)
14. Rotor full load kw: 200 × 0.01 = 2kw
    Assume 1 hp = 1 kva and 1% for rotor
15. Rotor full load amperes: kw/volts = amp 2,000/125 = 16 amp
    Assume 125 volts—standard
16. Conductor amperes: 16 × 1.25 = 20 amp
    Assume 125% and NEC 430-22
17. Wire size: No. 12 AWG at 75°C
    NEC Table 310-12
18. Conduit size: two No. 12 in ¾-in.
    Trade practice

This, then, completes the design calculations of the wiring requirements for the 200-hp 440-volt synchronous motor. Lines 6, 7, 8, 11, and 12 should be discussed with the manufacturer to obtain the best controller for the least money. Unless the designer is also the chief engineer, further information will probably be provided by the electrical supervisor, who is sometimes aware of criteria not previously divulged to the designer. This will then permit a more exact specification of equipment.

4.16 The Exciter

The excitation system supplies dc power to the field of a synchronous motor. The types of apparatus for supplying this dc power are numerous. In some cases, a dc generator is mounted on the shaft of the motor and feeds its own field. In other cases, the excitation system is located remotely from the motor. In this case, conductors have to be run between the motor and the excitation source. We will not discuss the function of the excitation system at this point.

Our immediate concern is how to size the conduit and conductors. In most cases, the designer is informed that a synchronous motor of a certain size will be provided. This is probably all the information the designer will receive during the early stages of design. A word of caution at this point: Remember that if we have not received instructions giving specific sizes of equipment, then we are only guessing. In many cases, this is the only information available, and when construction schedules are to be met, the calculated guess of the designers has to be good enough. What do we do in such a situation? As outlined previously, consider all the known facts, then make a decision based on these facts. Do the design on the drawing and mark with a "Hold." When the drawings are checked by the supervisory and construction personnel, these points will be questioned, and by that time the correct information may be available.

4.17 Size Evaluation and Short-circuit Ratio

With excitation systems, we nearly always have to make an educated guess. The only correct way of obtaining current ratings is to wait until the equipment is purchased. The manufacturer is then contacted, and the correct information is obtained. In practice, this is usually too late. We will establish a rule of thumb that will at least "keep us in the ballpark" and enable the designer to proceed without interruption.

The field current supplied by the exciter is dependent on a number of factors. One is the short-circuit ratio of the machine. This can be anywhere between 0.60 and 1.40. Second, the current density in the conductors will vary from 1,500 to 2,500 amp/sq in. The addition of losses and voltage drop must also be considered. These factors, along with the physical limitation of how many turns can be wound on a pole, preclude the possibility of accurately second-guessing the actual current required for excitation for a specific machine. It will vary with the manufacturer.

For our purposes, we will assume that the power required for excitation of a synchronous motor, at maximum, will be 1 to 2%, depending on power factor.[1] For unity power factor, we will use 1%, and, as the power factor decreases, we will multiply by the reciprocal of the new power factor. This will then increase the size accordingly. The voltage for excitation system is usually standard at 125 volts up to 10,000 kw and 250 volts for larger sizes. For a synchronous motor, the rating

[1] See Sec. 7.2.

will be in horsepower. We will consider 1 hp equivalent to 1 kva at 100% PF for purposes of evaluation.

4.18 Conductor Size

Now, we consider the *actual design* figures on our 200-hp field and exciter. We can then compare with our calculated guess (see Sec. 4.15) and determine the difference.

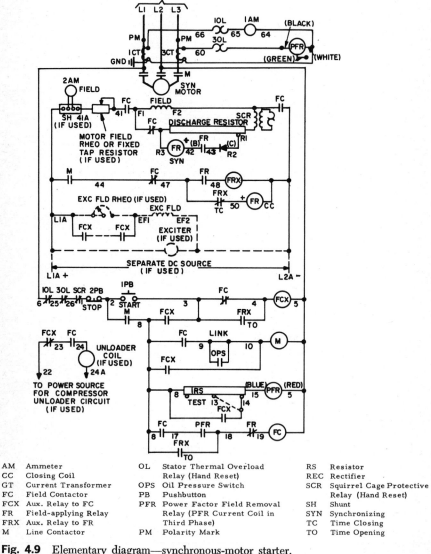

AM	Ammeter	OL	Stator Thermal Overload Relay (Hand Reset)
CC	Closing Coil		
GT	Current Transformer	OPS	Oil Pressure Switch
FC	Field Contactor	PB	Pushbutton
FCX	Aux. Relay to FC	PFR	Power Factor Field Removal Relay (PFR Current Coil in Third Phase)
FR	Field-applying Relay		
FRX	Aux. Relay to FR		
M	Line Contactor	PM	Polarity Mark

RS	Resistor
REC	Rectifier
SCR	Squirrel Cage Protective Relay (Hand Reset)
SH	Shunt
SYN	Synchronizing
TC	Time Closing
TO	Time Opening

Fig. 4.9 Elementary diagram—synchronous-motor starter.

Actual design	Design calculations
1. Horsepower, 200.................	200 hp
2. Full load amperes, 209.........	210 amp
3. Exciter volts, 125 volts dc.....	125 volts dc
4. Exciter kw, 1.71...............	2.0 kva at 100% PF = 2 kw
5. Maximum field amperes, 13.12...	16 amp

The result in line 5 is that we must use a No. 12 conductor, regardless of whether we use the actual design figures or the calculated design figures.

The short-circuit ratio of the above motor was 1.015. If this value is increased, then a necessary increase in field current will follow. The short-circuit ratio is a factor used in motor design that determines the effectiveness of the field winding. The higher the short-circuit ratio, the more copper is required in the field winding. As the short-circuit ratio increases, the stability and regulation quality increase. The Code is not specific regarding a *multiplier* for exciter conductors (see Sec. 4.15 of this book). We will use 125% where this figure is not specified. The diagram in Fig. 4.9 is not intended for study. It is shown here to point out the possible complexity of a synchronous motor starter and the advisability of contacting the manufacturer when purchasing this piece of equipment.

Chapter Five

Generators and Variable-speed Drives

5.1 Generators

Generators, like motors, are hard to classify as to exact type. The variations in design are innumerable. However, the basic groups can be described. These include the ac-3 and ac-1 phase, and the dc-shunt, dc-series, and dc-compound. In each case, we see that the identification is similar to the motor designation. With minor design modifications, a motor can be made a generator by simply attaching a "prime mover" (an engine) and driving the motor. It follows then that the wiring associated with a generator should be similar to the wiring for a motor. This is true up to a point. Compliance with the Code will allow certain considerations. The protective-device approach is also somewhat different. We will consider these factors as we treat each type of generator.

5.2 The AC Generator—Three-phase

The ac generator consists of a stator similar to the squirrel-cage stator. Three or four conductors are brought out from the stator to the appropriate switchgear. These leads are phases a, b, c (and neutral). The stator winding can be connected in star or delta. If it is connected in delta, then the fourth conductor will not be necessary. In regard to only the star-connected generator, we will point out that the neutral will probably be grounded in a four-wire system. Therefore, under NEC 240-12, "No overcurrent device shall be placed in any permanently-grounded conductor" except under the conditions outlined by NEC 240-12 and also according to the manufacturer's instructions.

The rotor of an ac generator is usually a *salient-pole* type. By definition, *salient* means *projecting out;* therefore, we have a rotor whose poles are projecting out from the center. The number of poles is dependent on the operating speed and frequency required. The poles are supplied with dc power through two slip rings and brush gear.[1] The dc supply can be a small generator mounted on the generator shaft, or it can be a remotely located dc supply. Being dc, the polarity of the poles and the number of poles are constant. Therefore, the speed of the rotor relative to the stator determines the frequency of the output voltage.

$$f = \frac{\text{rpm} \times \text{poles}}{120}$$

The reader can study the theory of magnetic fields and waveshapes in any general physics book. Therefore, we will not consider the theory of operation. Our only concern here is the necessary wiring and the associated control equipment required for the generator.

5.3 Generator Ratings

In the case of a generator, we must think in reverse; i.e., the conductors are supplying power to the overcurrent devices in the *power-distribution center.* As previously explained, a three-phase star-connected ac generator will have four conductors coming out of the stator.

The section of the Code for generators is Article 445[2] and the section covering the size of conductors is Article 445-5. We notice in Article 445-2 that the rating of a generator must be given in kva or kw at normal volts, amperes, and rpm. Usually, the power factor is also given. This is the estimated *load* power factor. In other words, the generator will supply its fully rated output in kw when the power factor of the load is as specified (or higher) on the nameplate. The conductors from the generator terminals to the supplied equipment shall have a current capacity of 115% of the nameplate current rating of the generator. In the case of a four-wire star-connected generator, the neutral conductor must be the same size as the phase conductors supplying

---

[1] Most small generators are now the "brushless" type, i.e., no slip rings.
[2] As large generators usually belong to utility companies—see NEC 90-2(b)(5).

the equipment. The generator rated in kva requires a simple calculation to determine the full load current. For three-phase power, this calculation is as follows:

$$\text{Power} = \text{va} \times \text{power factor} \times \sqrt{3}$$

or

$$\text{Watts} = \text{va} \times \cos\theta \times 1.73$$

where $\cos\theta$ is power factor, expressed as a decimal.

NOTE: *It may seem unnecessary to explain that $\cos\theta$ is to be expressed as a decimal. In fact, that is the only way $\cos\theta$ can be expressed. However, in actual practice, the power factor is sometimes expressed as a percentage. For example, if $\cos\theta$ is 0.80, then the expression normally used is, The power factor is 80%.*

5.4 Power Factor

Now, we must remember that the above equation represents power actually available to be used by the load. The power factor in an industrial plant is dynamic. It is constantly varying. When specifying the value for power factor, a calculated guess is made or an average condition is assumed based on experience and/or familiarity with plant loads. Power factor is represented by a number *between 0 and 1*. In actual practice, an attempt is made to maintain the power factor as close to unity as possible, but usually it is between 0.6 and 0.95.

5.5 Power

There are two kinds of power required by an industrial plant: power to supply the equipment ($EI \cos\theta$; this is *true* power) and reactive power ($EI \sin\theta$) to supply the losses due to magnetic fields in the electric equipment. Both types of power are transmitted over the same conductors. However, only the true power is measured by the power-company energy (watt-hour) meters.

The output of an ac generator must be adequate to supply both the normal power and the reactive power of the load. Consider an ideal load of 1,000 kva at 1 PF (unity power factor). Then

$$\text{Power (watts)} = 1{,}000 \text{ kva} \times 1$$
$$= 1{,}000 \text{ kw}$$

The generator will then have to be designed for an output of 1,000 kw. Now, if the power factor of a 1,000-kva load is 0.80, then

$$\text{Power (watts)} = 1,000 \text{ kva} \times 0.8$$
$$= 800 \text{ kw}$$

This means that only 80% of the same 1,000-kva output is available to drive machinery. Consider a large industrial plant with numerous motors and inductive-type loads, and with an 80% power factor. This means that one-fifth the total generator power used by the plant is for nonproductive purposes. We see then that it is an advantage to maintain a high power factor. The method and calculations involved in power-factor correction will be covered in a later chapter. Our only concern at this point is to understand the rating of a generator. We can then calculate a realistic full load current when nameplate information is not available.

5.6 Generator Output

Consider a typical generator rating: 1,000 kw, 0.8 PF, 480 volts, three-phase, 60 cps, star-connected, four-wire. The power factor specified is the assumed power factor of the load that the generator has to supply. Now, we must be very careful here to determine the amount of power actually required. If we say we require actual productive power, then we mean that kilowatts are required. There will have to be some reactive power requirements; therefore, this must also be supplied by the generator. This means that we require more output than the 1,000 kw. Now, the nameplate rating is 1,000 kw at 80% PF. To interpret this, we will consider what the manufacturer is trying to tell us with the nameplate. He is saying, *If the plant power factor is 0.8, then I will provide a generator that will supply the reactive power and still give you 1,000 kw for usable power.* Now, when we consider the full load current of the generator, we must consider it on a kva basis:

$$\text{Generator output in kva} = \frac{1,000 \text{ kw}}{0.8 \text{ PF}} = 1,250 \text{ kva}$$

Because the reactive power utilizes the same conductors as the active power, they must be sized to carry this extra power. We also see that the power we must consider is not the 1,000 kw on the nameplate, but 1,250 kva, as calculated. To calculate the full load amperes is simple; we use the following equations:

$$I = \frac{kva}{kv} \qquad \text{for single-phase}$$

$$I = \frac{kva}{\sqrt{3}\,kv} \qquad \text{for three-phase}$$

The full load current of the generator used in our example would be

$$I = \frac{1{,}250 \times 10^3}{1.73 \times 480} \qquad \text{or} \qquad \frac{1{,}250}{1.73 \times 0.480}$$

$$= 1{,}505 \text{ amp}$$

5.7 Generator Conductors

The Code specifies that conductors between a generator and the supplied equipment must be rated at not less than 115% of the nameplate current rating of the generator (NEC 445-5). In the early design stages of a project, the equipment may not have been ordered. Therefore, the nameplate will not be available, in which case we follow the previous procedure to determine the full load current.

The conductors required to meet the code will be

$$1{,}505 \times 1.15 = 1{,}730 \text{ amp}$$

When we get into large current ratings, a single conductor sometimes becomes impractical, in which case we use smaller conductors in parallel. We can use two or more; the selection should be based on using standard or reasonable sized conduit and wire. For our purposes, we will consider three conductors in parallel. This means that each conductor must be rated at minimum 577 amp per conductor. The neutral must also be rated to carry 1,730 amp (NEC 445-5). Therefore, we use three conductors in parallel for the neutral. We now have 12 conductors between the generator and its supplied equipment. This requires three separate conduits. Here we must consider NEC 300-20. *To prevent inductive heating of any metal enclosures (rigid conduit for example), all phase conductors and the neutral shall be grouped together.* Therefore, each conduit will contain conductors of phases *a, b, c,* and *N*. From NEC Table 310-12, we see that 1,250 MCM at 75°C is rated at 590 amp. Also, from NEC Table 1, Note 2 we see that four 1,250-MCM conductors require a 6-in. conduit. Our complete power-conductor system then consists of three separate conduits, each containing four conductors. Now, we must also point out that NEC

Table 310-12 specifies "not more than three conductors per conduit or raceway." We have more than three conductors per raceway, but our fourth conductor is a neutral, and, from Note 10(a) of Notes to Tables, NEC 310-12 to 310-15, we see that this does not apply when the fourth conductor is a neutral.

5.8 The Exciter

The generator is essentially a synchronous motor, run in reverse, and also requiring dc power for the field winding. This is supplied by an exciter. Here again, in the early stages of design, we will probably encounter the problem of insufficient information. We must now make a calculated guess on the exciter rating, calculate the current, size the conductors and raceway, and complete our design by marking with "Hold," following our normal procedure for unconfirmed information. A generator requires more stability than a motor. As we pointed out before, this is dependent on the *short-circuit ratio* of the machine. In a generator, we must assume that the short-circuit ratio is at least 1.25. This will require a larger field current and consequently a larger rated exciter than the equivalent size in a synchronous motor. The size of the exciter will vary inversely with the speed (but not necessarily proportionately). For example, a 500-kva 900-rpm 80% PF generator requires approximately a 6-kw exciter capacity. By comparison, a 500-kva 200-rpm 80% PF generator requires approximately a 15-kw exciter. In the first case, the exciter is just over 1% (1.2%) of the generator rating. In the second case, the exciter is approximately 3% of the generator rating. This leads to the conclusion that in order to make a calculated guess at the exciter rating, we must also guess at the speed of the generator. The probability of guessing the speed, short-circuit ratio, and exciter size correctly is remote. We must, however, make some decision so that we can continue with the design. We will establish a specific ratio of generator rating to exciter rating at 3% of output kva. Remember that the output kw should be divided by the power-factor rating of the generator to obtain the kva. The power factor must be assumed at 80%, unless otherwise specified. We will establish the voltage of the exciter at 125 volts dc, up to a 10,000 kva generator rating. Above that, we will assume a 250 volt dc rating. By establishing a routine constant, we know that we will be reasonably close and probably slightly oversize. The inaccuracy in our calculated guess will probably be neutralized by the *gaps* between standard wire size. The difference between a No. 12 at 20 amp and a No. 10 at 30 amp is a 50% increase, which is probably a far greater spread than *our* error would be. The conductors and/or raceway will be marked "Hold" in any event and the "Hold" will only be removed when the correct information is received. Alternatively, someone in authority may accept the calculated guess in lieu of waiting for the actual information and remove the "Hold."

5.9 Sample Calculation

We will now consider a sample calculation of the generator and exciter previously mentioned.

1. 1,000 kw, 80% PF, 480 volts, three-phase, 60 cps, star, four-wire — Original data and assumed PF
2. Full kva output: $1,000/0.8 = 1,250$ kva — Assumed 80% PF
3. Full load current: $(1,250 \times 10^3)/(1.73 \times 480) = 1,505$ amp — At 480 volts—3-phase
4. Wire size: $1,505 \times 1.15 = 1,730$ amp — NEC 445-5
5. Use three in parallel; $1,730/3 = 577$ amp each — NEC 300-20
6. 590 amp = 1,250 MCM at 75°C — NEC Table 310-12
7. Neutral capacity: 1,730 amp; use three 1,250 MCM — NEC Tables 310-12 and 445-5
8. Conduit: rigid steel; use 6-in. for four 1,250 MCM — NEC Table 1, Note 2
9. Exciter: $1,250 \times 0.03 = 37.50$ kw (Hold) — Assume 3%
10. Field amperes maximum: $37.5/125 \times 10^{-3} = 300$ — Assume 125 volts
11. Conductor size field—$300 \times 1.15 = 345$ amp — NEC 445-5
12. 380 amp = 500 MCM at 75°C (Hold) — NEC Table 310-12
13. Conduit: two 500 MCM; use 3-in. rigid steel — NEC Table 3A

Note that no speed is given or assumed. We must assume the worst case, but this was allowed for in the 3% constant. We now refer to a text[1] giving approximate exciter capacities for 60-cps 80% PF salient-pole generators. The family of curves indicates that for 100 rpm and 1,000 kva, a 32.5-kw exciter would be required. For 900 rpm and 1,000 kva, a 7.5-kw exciter is required. In the first case 350 MCM THW is required in a 2½-in. conduit. In the second case, a 7.5-kw exciter requires a 69-amp conductor No. 4 THW and 1-in. (below-grade) conduit. We see then that there must be an evaluation of cost against time and material. If the designer has time to wait for accurate information, and construction can be held up pending the decision, then the obvious conclusion is to wait. If the construction is proceeding as drawings are being made, then it is false economy to wait.

5.10 Cost Evaluation

Assume that the considerations are the difference in cost between approximately 30 to 50 ft of 2½-in. with two 350-MCM and 30 to 50

[1] J. H. Kuhlmann, "Design of Electrical Apparatus," 2d ed., p. 247, John Wiley & Sons, Inc., New York, 1949.

ft of 1-in. with two No. 4 AWG. The labor involved in the installation is nearly the same. Therefore, the only real cost difference is material. If we did not know the exact difference in cost, and assumed a difference of $1 per foot, under the worst conditions in our outline, we would be considering $50. The designer's time should be worth at least $10 per hour, including overhead. Therefore, it will cost $10 even to consider the problem and work out the cost differential. If there is a delay in the construction, it could mean a delay in pouring concrete involving a number of men. This could be because the drawing is not ready, due to the hold on the conduit for the exciter. Down the line people make inquiries about the drawing; when all this time is evaluated, the $50 seems small in comparison to the inconvenience and delays. Therefore, look beyond the immediate problem when making these decisions. Considering all aspects, oversizing is probably cheaper than spending the time making time-consuming evaluations in the exciter problems.

5.11 DC Generators

Dc generators are not normally used for supplying ordinary general-purpose power. The application of dc generators is nearly always a specific application, such as supplying dc power for variable-speed drives in paper mills, steel mills, and similar operations. The usual procedure in selecting this type of system is to contact the various manufacturers and discuss the application with them. They will supply a complete system, with drawings and wiring instructions. In the majority of systems, the dc generator will be driven by an ac motor, probably of the synchronous type. In this case, the designer usually has no alternative but to wait for the manufacturer's drawings. On receipt of these drawings, the designer will then show all the supplied equipment in the various locations on the drawings. He will then have to extract all the external wiring from the manufacturer's drawings and show it on the normal electrical drawings. This wiring will then fall into the category covered by the Code under branch-circuit conductors, feeders, generators, control, etc. The current and loads should be taken from the manufacturer's information, except where the Code specifically states that the tables should be used (see NEC 430-6 and 445-5).

5.12 The DC Generator Exciter

The exciter for a dc generator is not as simple to consider as the exciter for an ac generator. Sometimes the generator field is self-excited, or alternatively, it can be excited from a remote source. In either case, a rheostat in series with the field is usually used to control the field current. The type of excitation will only be known when the manufacturer supplies his drawings. As with the generator itself, the only solution is to wait until this information is received on *certified* drawings supplied by the manufacturer. Do not make the mistake of using drawings of proposed systems supplied by other manufacturers. They may not accurately represent the system that is ultimately purchased.

5.13 Variable-speed Drives

Variable-speed drive systems can be ac with variable-frequency generators supplying squirrel-cage motors, or they may be a specially designed form of synchronous motor. When the speed of the ac generator is varied, the speed of the motors will vary accordingly. The specific systems are all custom designed, but usually using standard equipment as manufactured by the vendor. The procedure for wiring this type of equipment is the same as outlined for the dc generator. The actual certified drawings must be obtained before the external wiring can be designed. A similar procedure is followed for variable-speed dc drives.

The components of this type of drive system are an ac motor driving a dc generator (this is known as an MG set). The dc generator then supplies power to the various motors. The speed is varied by controlling the output from the generator by means of an automatic regulator and feedback system. The manufacturer will supply the certified connection drawings, showing all the external wiring between components. The designer need not necessarily be familiar with the actual drive system, but he must be able to analyze a set of drawings accurately and be able to trace out the external wiring from the internal wiring. Sometimes it is not specifically called out as external wiring by the manufacturer.

The latest trend in variable-speed drives is to use an SCR (silicon controlled rectifier) system. The power to a dc motor is supplied from a normal ac source into a rectifier power supply. This power-supply unit has all the necessary controls internally. The installation simply means bringing in ac supply to the SCR power "pack"—taking output (that is, dc power) to the motor. Varying the speed simply requires turning a knob on the power pack.

5.14 Horsepower and Load

NEC 430-88 and 430-89 point out the protection against overspeed due to field weakening, etc. This is usually considered by the manufacturer in his design. The most important point for the designer to remember is that horsepower is a function of speed.

$$\text{hp} = \frac{2\pi NT}{33,000}$$

where N = rpm
T = torque, ft — lb.

When calculating the loads for sizing of conductors, remember that the highest horsepower is at the highest speed, and therefore the system conductors must be capable of handling the loads at the highest speeds. The *actual horsepower* ratings should be specified by the manufacturer with the full load amperes at maximum horsepower and efficiency and, in the case of an ac motor, also a function of power factor. Guessing efficiency and power factor at a particular speed is virtually impossible when the only information available is the horsepower rating.

5.15 Motor-starting Characteristics

Particular attention should be paid to the various starting methods and motor characteristics related to starting. This is necessary when considering overcurrent devices. Table 5.1 shows the various inrush values.

TABLE 5.1 Approximate Full-voltage Starting-inrush Factors for AC Motors

Type of motor	Inrush current, approx %
Induction motor—squirrel cage	600
Wound rotor—with full resistance	150
Synchronous motor:	
Low-speed low-torque	300
Low-speed high-torque	550
High-speed low-torque	400
High-speed high-torque	550

NOTE 1. High speed is over 500 rpm
NOTE 2. Multiply values in NEC Table 430-150 by the above percentages to obtain approximate inrush current except where NEC Table 430-151 takes precedence.

The data in Fig. 5.1 show connections and characteristic curves due to specific starting methods.

Figures 5.2 and 5.3 show curves for starting large motors from a single generator or transformer.

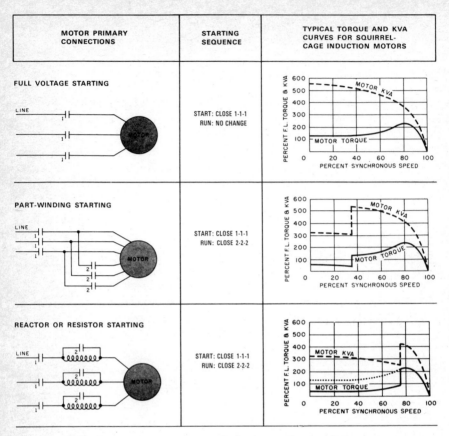

Fig. 5.1 Induction motors—starting methods, connections, and torque curves.

Generators and Variable-speed Drives 55

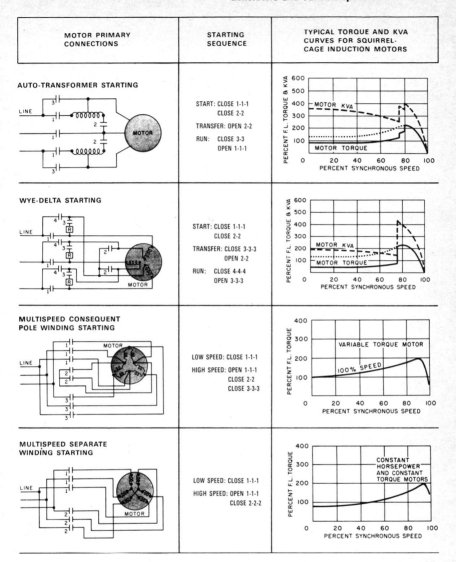

Fig. 5.1 (continued) Induction motors—starting methods, connections, and torque curves.

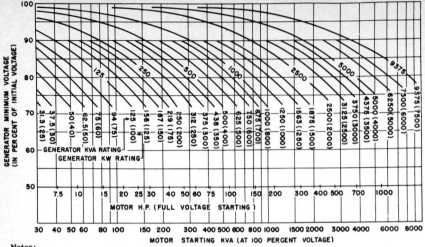

Notes:

(1) Scale of motor-hp is based on starting current being equal to approximately 5.5 times normal.

(2) If there is no initial load, voltage regulator will restore voltage to 100 per cent after dip to values given by curves.

(3) Initial load, if any, is assumed to be constant-current type.

(4) Generator characteristics assumed as follows:
   a. Generators rated 1000 kva or less
   Performance factor, K = 1.0
   Transient reactance, X'_d = 25 percent
   Synchronous reactance, X_d = 120 per cent
   b. Generators rated above 1000 kva

Fig. 5.2 Starting large motors on the generator.

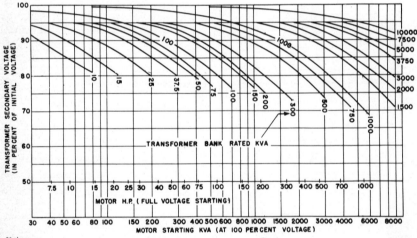

Notes:

1. Scale of motor hp based on starting current being equal to approximately 5.5 times normal.

2. Short-circuit kva of primary supply is assumed to be as follows:

Bank Kva	Primary Short-circuit Kva
10-300	25,000
500-1000	50,000
1500-3000	100,000
3760-10000	250,000

3. Transformer impedances are assumed to be as follows:

Bank Kva	Bank Impedance
10-50	3%
75-150	4%
200-500	5%
750-2000	5.5%
3000-10000	6.0%

4. Representative values of primary system voltage drop as a fraction of total drop are as follows, for the assumed conditions.

Bank Kva	System Drop/Total Drop
100	0.09
1000	.25
10000	.44

Fig. 5.3 Transformer voltage drops due to motor starting.

Chapter Six

Transformers

6.1 Transformer Function

A *transformer* is a simple static device used for raising or lowering the voltage in an ac circuit.

An obvious question is, Why should we want to lower or raise the voltage? The answer is very simple. *By doubling the voltage of a power-distribution or transmission line circuit, we can go 4 times the distance for the same amount of losses.* Also, instrument transformers reduce a hazardous voltage level to a lower safe level.

6.2 Transformer Principles

We will briefly cover the working of a transformer with the understanding that there are many variations on a theme. The basic principle, however, is applicable to all transformers.

Consider a loop of laminated sheet steel in rectangular shape and wound with wire at opposite ends. With a 2:1 turns ratio, the difference between input and output voltages would be in the same ratio, or 2:1. Here then we have the basis of design for all transformers. The significant factor to be considered is the turns ratio between the input and the output. The input voltage is termed the primary voltage E_1, and the output voltage is noted as the secondary voltage E_2. If we

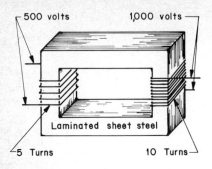

Fig. 6.1 Transformer with 2:1 ratio.

identify the primary and secondary turns as N_1 and N_2, respectively, we can establish the following equation: $E_1/E_2 = N_1/N_2$ for the total induced voltage in each winding. We can see by simple manipulations that N_2, for instance, is equal to N_1E_2/E_1.

The EMF induced in the second winding is proportional to *flux, frequency,* and the *number of turns.* This is shown in the following equation for induced voltage which is used in transformer design.

$$E = 4.44\phi_{max}fN10^{-8}$$

where f = frequency
N = number of turns
ϕ_{max} = maximum flux in CGS units
4.44 = design constant

The current in the transformer, when fully loaded, is inversely proportional to the number of turns in the secondary and primary windings.

$$\frac{I_1}{I_2} = \frac{N_2}{N_1}$$

I_2 would be equal to I_1N_1/N_2. Therefore $E_2/E_1 = N_2/N_1 = I_1/I_2$.

6.3 Transformer Selection

The preceding information is just an introduction to the main principles governing the design of a transformer. There are many more considerations confronting the transformer designer before an economical design is decided upon. This, however, is not a problem that concerns the average plant design engineer. The problems confronting him are a completely different set of problems. There must be thousands of different transformers available to the plant design engineer. His problem is to select the correct size and type at the correct price, also ensuring that delivery will be compatible with the particular project schedule.

Transformers have become highly standardized and usually the rating, weight, and size of different manufacturers' products are very comparable. The deciding factors are usually price and delivery. All power transformers are rated in va or kva, rather than true power, which is va cos θ. The impedance Z of the transformer is usually standard and is rated in Z% (percent impedance), according to size and type. For purposes of short-circuit calculations, this information is usually accurate enough when selected from any standard catalog.

6.4 The Autotransformer

Up to this point, we have discussed the conventional two-winding transformer. One of the variations of this is the autotransformer. Instead of two separate windings, only a single winding is used. The core material is the same as for the two-winding transformer. The single winding is placed on a laminated metal core, and a tap is selected which gives the turns ratio required. By using a common winding, space and copper are saved. However, the disadvantage is that there is a direct connection between the high-voltage and low-voltage winding.

The type of transformer illustrated in Fig. 6.2 is quite common in utility substations, where a high-voltage transmission line may be stepped down from 440 to 220 kv. Both primary and secondary are high voltage. Therefore, the problem of direct connection between primary and secondary is no longer important, because the same safety precautions are required for 220 kv as for 440 kv.

6.5 Transformer Rating

The autotransformer is rated in kva, like any other transformer. However, by using a common winding, some of the secondary load is carried

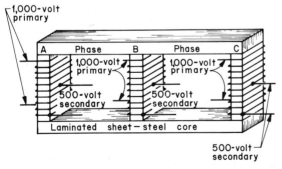

Fig. 6.2 Three-phase stepdown autotransformer with 2:1 ratio (phase connections not shown).

by the primary copper. If we have a load which requires a 1,000-kva three-phase transformer with a 2:1 transformation ratio, we would provide a two-winding transformer, rated at 1,000 kva. Alternatively, the equivalent autotransformer, smaller in actual size, would provide the same kva transformation capacity.

The equation to determine the equivalent autotransformer size is

$$kva_{1w} = kva_{2w} \frac{N-1}{N}$$

where N = HV/LV or turns ratio; HV/LV = high voltage/low voltage
kva_{1w} = kva, one-winding transformer
kva_{2w} = kva, two-winding transformer

To expand the equation, allowing the use of voltages instead of turns ratio, we have

$$kva_{1w} = kva_{2w} \frac{HV/LV - 1}{HV/LV}$$

$$= kva_{2w} \frac{(HV/LV - 1)LV}{HV}$$

$$= kva_{2w} \frac{LV(HV/LV) - LV}{HV}$$

$$= kva_{2w} \frac{HV - LV}{HV}$$

Considering an equivalent size autotransformer to replace the 1,000-kva two-winding transformer, we have

$$kva_{1w} = kva_{2w} \frac{N-1}{N}$$

$$= 1,000 \frac{2-1}{2}$$

$$= 1,000 \times 0.5 = 500$$

If the voltages were 460 to 230 kv we would have

$$kva_{1w} = 1,000 \frac{460 - 230}{460}$$

$$= 1,000(1 - 0.5)$$

$$= 1,000 \times 0.5 = 500$$

The autotransformer nameplate may show 1,000 kva capability, but the copper content and physical size will be closer to those of a two-winding 500-kva transformer. This would also partially reduce the cost of the transformer by the amount of material saved.

6.6 Autotransformer Limitations

The National Electrical Code places certain restrictions on the use of autotransformers. Article 200-4 specifies that autotransformers shall not be used on branch circuits, unless "the system supplied has an identified, grounded conductor, which is solidly connected to a similar identified grounded conductor of the system supplying the autotransformer."

In NEC 430-82(b), we find that the autotransformer can be used for motor starting under certain conditions.

NEC 410-76 deals with the use of the autotransformer as a ballast unit for lighting fixtures. Here again, certain conditions are attached to its use.

The rule to follow, then, is that all normal transformer requirements will be for the two-winding type of transformer with the autotransformer reserved for specific applications.

6.7 Power Transformers

Transformers with a rating larger than 500 kva are classed as power transformers. They may be single-phase, three-phase, two-winding, or autotransformer types.

The difference between the single-phase and three-phase transformer is that the latter consists of three single-phase units, using a common core and single enclosure. This we saw in the diagram of the autotransformer. There are advantages and disadvantages to both types. The obvious advantage of the three-phase transformer is the lower installation cost, plus the smaller amount of floor space it requires. The disadvantage is that a failure in one phase renders the transformer useless. Cost and availability are slight disadvantages, but these problems are offset by the advantages and should not influence the decision to any great extent.

6.8 The Single-phase Transformer

The single-phase transformer is very versatile in application. The disadvantage of this transformer is that three single units are required, along with the necessary floor space and the interconnecting wiring. This sometimes creates a space problem and also a higher installation cost. However, a failure in one transformer still allows the other two to function at 58% of the previous capacity of the three. This requires a reconnection to open delta, which will be discussed later. This advantage, along with the fact that one spare single-phase transformer will, in all probability, be adequate to maintain numerous three-phase banks, requires careful consideration when the three-phase or three single-phase units are selected.

6.9 Distribution Transformers

Distribution transformers are rated from 3 up to 500 kva, the implication being that anything less than 500 kva would be used for a utilization of power, rather than as a main source of supply. The comments on power transformers in Sec. 6.7 are equally applicable to the distribution transformer. The terms *lighting transformer* and *power transformer* are sometimes used with reference to small distribution transformers. This is a distinguishing factor between three-phase 480-volt transformers supplying machinery loads, and three-phase 208/120-volt four-wire transformers feeding lighting panels.

These transformers may be the dry type, oil-filled, or filled with nonflammable liquid. The application and location will determine the selection.

6.10 Control Power Transformers

Control power transformers are usually small (up to 10 kva), single-phase, dry-type transformers. The function of these transformers is to provide a low-voltage source of power for control circuits. Usually, they are required to reduce a 480-volt circuit to 120 volts for motor control. They are not to be confused with the instrument-type potential transformer.

6.11 Potential Transformers

A potential transformer is a transformer designed specifically to reduce a high voltage to a low voltage for purpose of instrumentation. The accuracy of these transformers must be maintained within certain limits with a specific load burden. If the load burden is exceeded then the accuracy of the transformer is affected. These transformers may be oil-filled for the high-voltage types or the dry type for the lower-voltage systems.

6.12 Isolating Transformer

In some cases, a problem arises with grounding of instrumentation systems. It is then necessary to isolate one part of the system from the other. This is usually accomplished by the use of a two-winding transformer with a 1:1 ratio, such as 120-volt primary and 120-volt secondary.

Because the two windings are only connected by a magnetic field, ground isolation can be obtained.

6.13 Current Transformers

The current transformer can be considered the partner of the potential transformer. Its function is to produce a voltage proportional to the line current with predictable accuracy. The accuracy is affected by the load burden, and since the function of the CT is to provide instrumentation with meaningful signals, the load burden should not be exceeded. The current transformer has an extremely high number of turns on the secondary relative to the primary. The sudden opening of the secondary under load may cause a transient discharge, which could produce insulation stresses exceeding the design limitations of the transformer. The load burden from a CT should be removed by first shorting the secondary terminals and then disconnecting the burden.

A current transformer is usually rated with the secondary fixed at 5 amp maximum loading. This means that the primary current will vary, e.g.,

(Primary) 1,500:5 (Secondary), or 1,000:5, or 500:5 amp, etc.

6.14 The Zigzag or Grounding Transformer

The "grounding transformer" designation is slightly misleading. By our earlier definitions a transformer is used to raise or lower voltages.

Its function is to obtain a neutral point from an ungrounded system. With a neutral being available the system may then be grounded. When the system is grounded through the zigzag transformer its sole function is to pass ground current.

A zigzag transformer is essentially six impedances connected in a zigzag configuration. The fact that transformer coils, core, and tank are used is a matter of using available parts.

The operation of a zigzag transformer is slightly different from that of the conventional transformer. We will consider current rather than voltage. It is true that a voltage rating is necessary but this is line voltage and is not transformed. It provides only exciting current for the core. The dynamic portion of the zigzag grounding system is the fault current. We must also "view" the system "backward"; i.e., the fault current will flow into the transformer through the neutral. In Fig. 6.3a and b we show the current distribution.

Without getting into a discussion of symmetrical components we will point out that zero sequence currents are all in phase in each line;

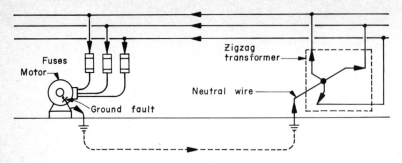

(a) Fault current path, three-line diagram

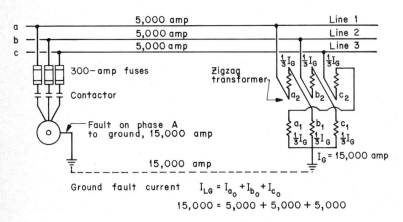

Ground fault current $\quad I_{LG} = I_{a_0} + I_{b_0} + I_{c_0}$
$15{,}000 = 5{,}000 + 5{,}000 + 5{,}000$

(b) Fault current path, three-line diagram

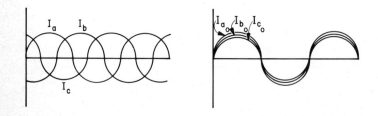

(c) Three-phase currents

(d) Three-phase zero sequence currents

Fig. 6.3

i.e., they all hit the peak at the same time. This accounts for the sum $I_{a_0} + I_{b_0} + I_{c_0}$ rather than phase currents which peak at intervals 120° apart. We see then that the current leaves the motor, goes to ground, flows up the neutral, and splits three ways. It then flows back down the line to the motor through the fuses which then "blow" or open—shutting down the motor.

The neutral conductor will carry full fault current and must be sized accordingly. It is also time rated (0–60 sec) and can therefore be reduced in size. This should be coordinated with the manufacturer's time/current curves for the fuse. It should also be remembered that a fault calculation may include a contribution that is not always present, and so when figuring time coordination, this should be taken into account.

To specify a zigzag grounding transformer proceed as follows:

1. Figure system line to ground asymmetrical fault current.
2. If relaying is present consider reducing fault current by installing a resistor in the neutral.
3. If fuses or circuit breakers are the protective device you may need all the fault current to open the fuse or the circuit breaker fast.
4. Obtain time/current curves of relay, fuses, or circuit breakers.
5. Specify zigzag for
    a. Fault current—line-to-ground
    b. Line-to-line voltage
    c. Duration of fault (determine from time/current curves)
    d. Impedance per phase at 100%; for any other, contact manufacturer

For estimating approximate equivalent three-phase size, use the following formula based on 10-sec rating.

$$\frac{kv}{1.73} (I_{sc})(0.064) = kva$$

NOTE: *An approximation to check formal calculations can be derived by assuming line current as unbalanced neutral current, and multiplying by 20 for 10-sec rating.*

6.15 Constant-current Transformer

This type of transformer is usually limited to street-lighting circuits or special circuits such as electric welding and sophisticated control circuits. The street-lighting transformer is a two-winding transformer, with one coil movable with respect to the other. By means of a counterweight and electromagnetic forces, a relative position is maintained that produces constant current from the transformer.

6.16 Transformer Overcurrent Protection

NEC 450 is the section dealing with the requirements for an adequate transformer installation. The whole section requires careful study to determine alternative approaches to installations. We will, however, discuss NEC 450-3 in detail.

A transformer is a relatively expensive item when compared to the overload device capable of protecting it. There are a number of ways of protecting a transformer, with varying degrees of sophistication. The method used depends to a great extent on the size and cost of the transformer. Consideration is also given to the function and reliability expected from the transformer.

Considering only liquid-filled transformers, we can place a protective device in either the primary or the secondary side, NEC 450-3(a)(1) and (2). Considering the primary side first, the overcurrent device must be set or rated at not more than 250% of the rated primary current of the transformer. This may be a device specifically for transformer

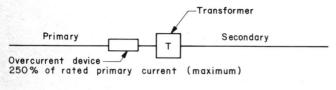

(a) NEC 450 – 3(a)(1)

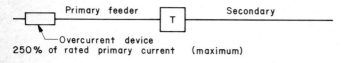

(b) NEC 450 – 3(a)(1)

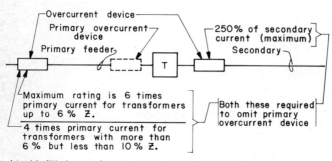

Liquid-filled-transformer overcurrent protection
(c) NEC 450 – 3(a)(2)

Fig. 6.4 Transformer overcurrent protection—alternative locations.

protection, or it may be the transformer "feeder" protective device, if the latter is rated within the 250% requirement.

An alternative to primary protection is allowed in NEC 450-3(a)(2), which states in effect that if the transformer is not self-protecting by the manufacturer's installation of a coordinated thermal overload, then we may install the overcurrent device in the secondary circuit. The device must be rated or set at not more than 250% of the rated secondary current, *and* the overcurrent device protecting the *primary feeder* must also be rated *or* set at not more than six times rated primary current. This applies only to transformers having up to 6% impedance. Special transformers having an impedance of 6 to 10% require the primary overcurrent protective device to be rated at no more than 4 times the rated primary current.

Figure 6.4 shows the alternative locations for overcurrent devices. NOTE: *NEC 450-3(b)(1) and (2) for dry-type transformers are essentially the same as for NEC 450-3(a)(1) and (2) except reading 125% instead of 250%.*

6.17 Differential Protection

A fuse or circuit breaker is dependent on thermal buildup to open the circuit. This takes time. The time may seem insignificant when we speak of fractions of a second. However, when we consider that at 60 cycles the current rises from zero to maximum, twice in $\frac{1}{60}$ sec, we can see that it bears consideration.

When a large expensive transformer is installed, it is usually protected by fast-acting relays (these will be discussed in detail later). This is specifically known as *differential relay protection*. The relay is operated by current transformers installed on both the input and output sides of the transformer. When a certain amount of current goes in, a specific amount should come out. The relays constantly monitor this relationship, and any significant change causes the relay to operate. This, in turn, trips the circuit breaker and opens the circuit to the transformer, hopefully before too much damage is done.

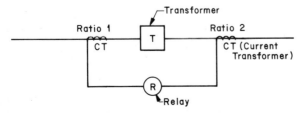

Fig. 6.5 Differential relay protection—single-line diagram.

6.18 Polarity

Transformer polarity is important when transformers are operated in a bank or in parallel. To indicate how the internal connections are made, the leads brought out of the case are numbered and lettered with specific notation. This has been standardized, and all high-voltage leads are numbered $H1$, $H2$, $H3$, etc., while the secondary leads are numbered $X0$, $X1$, $X2$, $X3$, etc. $X0$ indicates a neutral connection point.

6.19 Subtractive and Additive Polarity

Consider a single-phase transformer, with $H1$ and $H2$ terminals on the primary side. The secondary winding can be brought out ($X1$ and $X2$ matching the primary terminals) or it can be brought out with opposite terminals (arrangement $X2$ and $X1$), where $H1$ and $X1$ are diagonally opposite. The latter arrangement is additive polarity, and the former arrangement is subtractive. If the numbering is obliterated, then the $H1$ terminal is always the right-hand terminal when facing the high-voltage side of the transformer. The other terminals are numbered consecutively from right to left. To determine the polarity, connect the transformer as shown in Fig. 6.6. Additive polarity is standard for all single-phase transformers up to 200 kva, with high-voltage windings below 8,660 volts. All other single-phase transformers are connected with subtractive polarity.

6.20 Transformer Connections

When using three-phase power, we encounter the number $\sqrt{3}$ or 1.73. This number is the ratio by which the voltage and current differ in the two standard three-phase transformer connections. These are known as the wye (or star) and the delta. These two connections are always used in three-phase systems. The three coils of a three-phase

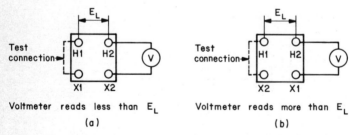

Fig. 6.6 (a) Subtractive polarity test connection; (b) additive polarity test connection.

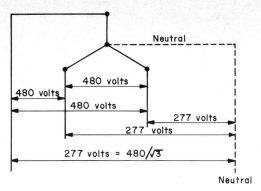

Fig. 6.7 Secondary star connection voltages.

transformer (or three single-phase) can be connected in the form of a Y or in the form of the Greek letter delta (Δ). The wye connection has a neutral point at the intersection of the three windings. Therefore, four wires will be available with this connection. The delta connection, not having a neutral point, only brings out three wires. Figure 6.7 shows the difference in voltage and current in the two connections, and it also shows that we obtain exactly the same power output from either connection.

NOTE: $E = 480$ volts for *phase* to *phase*.
 Then $E_N = 480/1.73$ for *phase* to *neutral*.
 $1.73 = \sqrt{3}$.
 Current in any *line* or *phase* is I.
 Three-phase apparent power is $E_{\text{phase}} I \sqrt{3} =$ va or $3 E_N I =$ va.

FOR THREE-PHASE EXAMPLE:
 480 volts and 100 amp; then kva $= 480 \times 100 \times 1.73 \times 10^{-3} = 83$ kva.
 Or 277 volts, 100 amp; then kva $= 3 \times 277 \times 100 \times 10^{-3} = 83$ kva.
 To obtain power (watts) is simply kva $\times \cos \theta$, where $\cos \theta =$ PF (power factor).

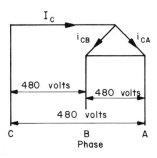

Fig. 6.8 Secondary delta connection voltages.

NOTE: $E = 480$ volts for *phase* to *phase*.
No neutral exists.
Current in any line is line current I.
But current in the *phase of the transformer* is $I/1.73$.

NOTE: I_c splits two ways and gives i_{ca} and i_{cb}, then $i_{ca} = I_c/1.73$.

Three-phase apparent power is $E_{\text{phase}} I \sqrt{3} = $ va.

FOR THREE-PHASE EXAMPLE:
480 volts and 100 amp; then kva = $480 \times 100 \times 1.73 \times 10^{-3} = 83$ kva.
Then star volts phase to neutral = deltavolts/1.73.
$$277 \text{ volts} = 480/1.73.$$
And delta volts phase to phase = star volts phase to neutral $\times$ 1.73
$$\therefore 480 \text{ volts} = 277 \times 1.73.$$

Figures 6.9 to 6.22 show standard transformer connections.

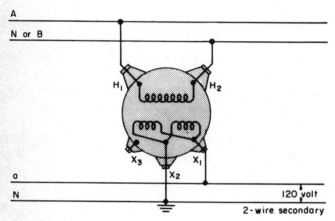

Fig. 6.9 Single-phase parallel connection.

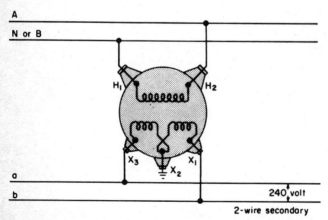

Fig. 6.10 Single-phase series connection.

Transformers 71

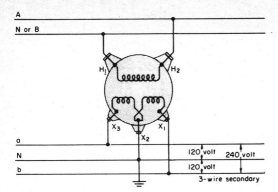

Fig. 6.11 Single-phase three-wire connection.

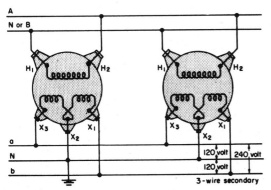

Fig. 6.12 Two single-phase connected in parallel.

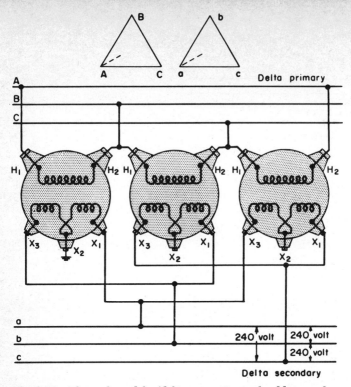

Fig. 6.13 Three-phase delta/delta connection with additive polarity and standard angular displacement.

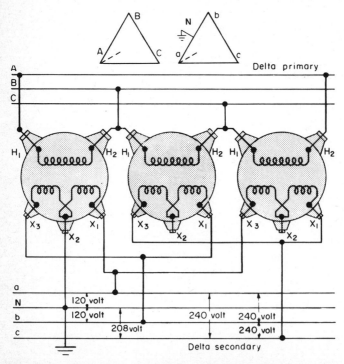

Fig. 6.14 Three-phase delta/delta connection with three-wire single-phase connection derived from one transformer. Additive polarity and standard angular displacement.

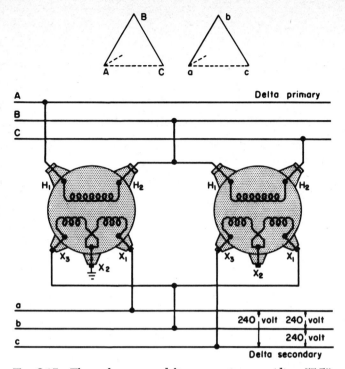

Fig. 6.15 Three-phase open-delta connection providing 57.7% kva of the three transformers of the same size connected delta/delta. Additive polarity and standard angular displacement.

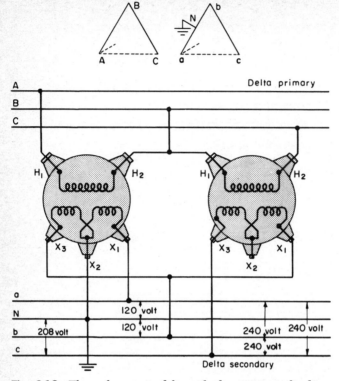

Fig. 6.16 Three-phase open-delta with three-wire single-phase connection derived from one transformer. Additive polarity and standard angular displacement.

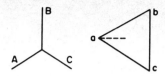
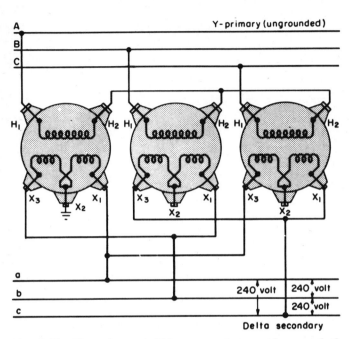

Fig. 6.17 Three-phase star/delta connection provides a method of changing from a 2,400-volt to a 4,160-volt system with the same transformer. Additive polarity and standard angular displacement.

76 Electrical Design Information

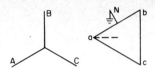

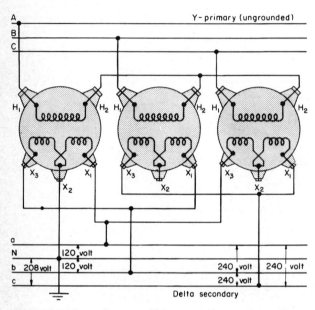

Fig. 6.18 Three-phase star/delta connection with single-phase available from the secondary connection (see Fig. 6.14). Additive polarity and standard angular displacement.

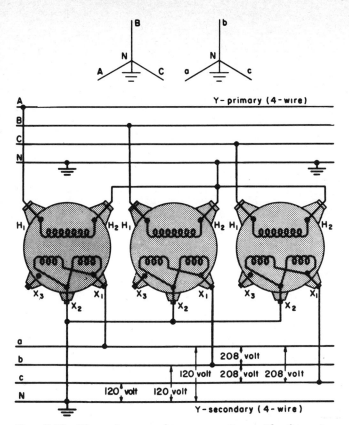

Fig. 6.19 Three-phase star/star connection with four-wire secondary connection. A problem with third harmonics can result with this connection. Additive polarity and standard angular displacement.

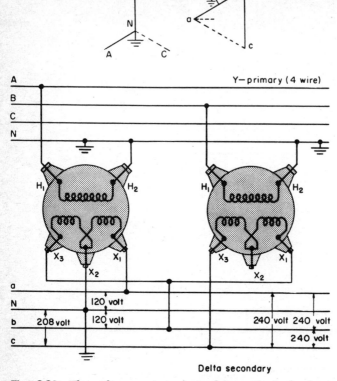

Fig. 6.20 Three-phase open-star/open-delta with single-phase three-wire connection derived from one transformer. Additive polarity and standard angular displacement.

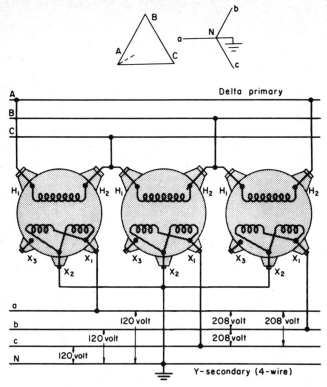

Fig. 6.21 Three-phase delta/star connection providing three-phase four-wire at the secondary. This allows for balancing single-phase loads. This connection also allows interconnection of multiple banks of transformers. Additive polarity and standard angular displacement.

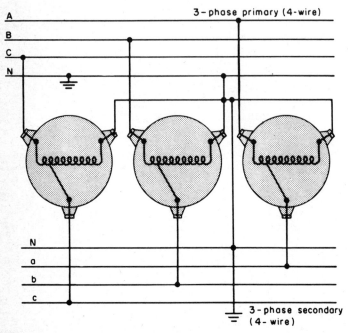

Fig. 6.22 Three-phase star/star autotransformer connection. Economic method of supplying 2,400-volt motors with 4,160-volt source. Also for high-voltage stepdown of transmission lines.

Chapter Seven

Power and Energy

7.1 Power

By definition, power is the time rate at which energy is transmitted. This is the usual definition. From a practical standpoint, when we talk of power, what do we mean? Usually, we have in mind a continuous source of energy that can be readily available at the location requiring motive power. To meet these requirements, we must have a source, a method of transmission, and a method of utilization.

When the age of electricity arrived, it was possible to take large quantities of power, transmit it over relatively small wires, and be able to use it with comparative simplicity. The distance of transmission is virtually limitless, and is controlled only by the economics associated with distance.

7.2 Power Measurement

In electric system analysis, power can be subdivided into various units. These units allow us to calculate and examine a particular section of a total block of power. We will consider apparent power (kva)[1] — active (real) power ($EI\cos\theta$ = watts) and reactive power ($EI\sin\theta$ = var)

---

[1] $K = 1,000$.

or kvar. We will consider alternating sinusoidal current and assume three-phase circuits as being balanced circuits.

7.3 Apparent Power

Apparent power is a scalar quantity with respect to single-phase and balanced three-phase systems. We will use the letter S to identify it. The quantity will be in va or kva. For single-phase calculations:

$$S = EI = \text{va}$$

For three-phase calculations:

$$S = \sqrt{3}\, V_L I_L$$

where line voltage V_L and line current I_L are used. An alternative is

$$S = 3 V_N I_L$$

where phase-to-neutral voltage and line current I_L are used.

7.4 Active Power

Active power is sometimes called *real power* and it will be referred to as such in this text. The unit is the watt and the quantity will be in watts or kilowatts. For single-phase calculations:

$$P = EI \cos\theta = \text{watts}$$

where E and I are effective values of voltage and current, and θ is the angle between the voltage and current vectors; $\cos\theta$ is the power factor.

7.5 Phase Angle

Here we will digress to explain the concept of *phase angle*. In an ac circuit, we have two continuous effects, which we do not have in dc circuits. These are inductance and capacitance. Inductance causes the current to lag the voltage by a specific time period, and the capacitance causes the current to lead the voltage by a specific time period. At this point, there is always some confusion as to how a voltage can get to a certain place before the current, or vice versa. The student usually tries to visualize how this occurs, which causes confusion and uncertainty. Here we will ignore the actual physics of motion, and we will consider the whole problem of "lead" and "lag" as a concept devised to enable us to represent in vector analysis this lead and lag time period.

7.6 Time Period

We know from our mathematics review in Part 2 of the book that it is quite simple to show an arrow and assume that it is rotating counterclockwise. This arrow could represent the voltage. If we say that it takes 1 sec to make a complete revolution, then we have already established a time period. Now, if we say that we will show a current lagging the voltage by ¼ sec, we could show another arrow 90° behind the first arrow. This would indicate a time lag of ¼ sec.

It should now be obvious that a time period can be represented by an angle. It should also be obvious that ½ sec would be 180°, and ¾-sec lag would be 270°. This, then, is the phase angle. When the current and voltage are together in time, then they are said to be *in phase*. When they are not together in time, they are said to be *out of phase*. We have used an arbitrary 1-sec time period for one full circle. We can also say *one full cycle* or *one cycle per second*.

7.7 Real Power

Now we will return to our equation for real power:

$$P = EI \cos \theta = \text{watts}$$

Here we are stating that the power in watts is equal to the product of the voltage and current and the cosine of the angle between the current and voltage. We recall from the chapter on trigonometry (Chap. 17) how to obtain the cosine of a given angle. For three-phase calculations:

$$P = \sqrt{3}\, EI \cos \theta = \text{watts}$$

where E and I are line-to-line voltage and line current, respectively.

7.8 Reactive Power

Reactive power is the portion of the power supply necessary to produce the magnetic flux for the various magnetizing fields. It is identified by the letter Q, and the quantity is expressed in var or kvar. For single-phase calculations:

$$Q = EI \sin \theta = \text{var}$$

where E and I are voltage and current, and θ is the angle of lead or lag between the voltage and current. For three-phase calculations:

$$Q = \sqrt{3}\, EI \sin \theta = \text{var}$$

where E and I are the line-to-line voltage and line current, respectively.

7.9 Power Addition

If we supply a system with a quantity of apparent power, i.e., va or so many kva, and this system contains equipment requiring magnetic flux, then this apparent power will be divided into two components. These are the *real component* and the *magnetizing* or *reactive component*. If we show this in complex or cartesian notation, we have

$$S = P + jQ = \text{va} \qquad \text{for sinusoidal balanced three-phase circuits}$$

7.10 Vector Addition

We also know, from our chapter on vector manipulation (Chap. 18), that the j is a 90° operator. Referring back to apparent power, we remember that it is a scalar quantity requiring magnitude only. Again from the mathematics review in Part 2, we remember that the magnitude of a complex number is obtained by finding the square root of the sum of the squares, or

$$S = \sqrt{P^2 + Q^2} = \text{va}$$

In other words, we add the real power and the reactive power vectorially to obtain the apparent power. We can see that by manipulating the various quantities, if we know any two, we can find the other one. To present these in the normal form, we have three basic equations:

$$\begin{aligned} S = \text{kva} &= \sqrt{(\text{kw})^2 + (\text{kvar})^2} \\ P = \text{kw} &= \sqrt{(\text{kva})^2 - (\text{kvar})^2} \\ Q = \text{kvar} &= \sqrt{(\text{kva})^2 - (\text{kw})^2} \end{aligned}$$

This is shown graphically in Fig. 7.1.

7.11 Ammeter Reading

The current indicated by an ammeter will read the vector sum of the real kw current and the reactive kvar current. This of course would indicate the current associated with the apparent power. Here we encounter a common design problem. The previous statement, that an ammeter will read both kw current and kvar current, immediately points

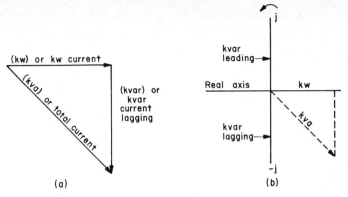

Fig. 7.1 (a) Power triangle; (b) alternative power triangle.

out that the conductors supplying a block of power will have to carry both currents (added vectorially) if the load contains any magnetic field devices. The reactive current contributes nothing to output power and yet it is necessary to supply the magnetic fields. What then is the procedure for analysis of this type of problem? Here we encounter the power factor $\cos \theta$, and the subject of power-factor correction.

7.12 Power Factor

In our previous digression, we had an introduction to *power factor*, i.e., $\cos \theta$. This factor is an extremely important quantity, both to the plant engineer and to the plant accounting department responsible for paying the electricity bill. It is very simply derived. It is a ratio of the real kw power to the apparent kva power.

$$\text{PF} = \cos \theta = \frac{\text{kw}}{\text{kva}}$$

By cross multiplication, it is simple to derive the other two equations:

$$\text{kw} = \text{kva} \cos \theta$$

$$\text{kva} = \frac{\text{kw}}{\cos \theta}$$

The power factor can be either leading or lagging, and it is indicated by a power-factor meter or by using voltmeter, ammeter, and kw meter and making the necessary calculations. It is not always obvious whether the leading or lagging current is going into the load or into the source of supply. Here is where an understanding of the various equipment characteristics is necessary, as well as an understanding of leading and lagging var terminology. The latter is simply another way of expressing the effect of reactive power on the system, instead of using the cos θ power factor.

7.13 Unity Power Factor

If we have a specific amount of *apparent power* (kva) supplying a *real power* (kw) load, the optimum condition would be at unity power factor or cos θ = 1. This would give the following equation:

$$1 \text{ kva} = 1 \text{ kw} \quad \text{at unity power factor}$$

All pure (noninductive) resistive-type loads will have a power factor of cos θ = 1. This would include loads such as heating elements, incandescent lighting, resistors, and any other noninductive noncapacitive device.

7.14 Lagging Power Factor

Any device having magnetic fields or magnetic coils will require *reactive power* or *var*. It will also cause the power factor to be less than unity and lagging. All induction motors, underexcited synchronous motors, uncorrected fluorescent lighting, reactors, ballasts, choke coils, mercury vapor lighting, induction heating, and other similar types of loads will require var. We can then state that a *lagging-power-factor* load will require var to go into the load. This means that the generator or source of supply will have to supply these var. As a power-company meter is measuring kw only, it is necessary for the power company to be reimbursed for var via various clauses or penalties outlined in the billing schedules. It is sometimes advantageous to have good (high) power factor and avoid penalties.

7.15 Oversize Motors

The average production plant requires numerous types and sizes of induction motors. The more heavily loaded the induction motor is, the better the power factor will be. The sizing of an induction motor to the load it is driving is important. Unfortunately, the sizing of a motor is usually determined by a mechanical engineer and is possibly oversized. Here, then, we encounter one source of poor power factor.

7.16 Oversize Exciter

A synchronous motor is sometimes used to drive a compressor in a production plant. In Chap. 4, we had an introduction to the synchronous motor and its exciter. It is characteristic of this type of motor that if it is underexcited, it will be similar to the induction motor and will require var and will contribute to poor power factor. As the excitation is increased, the power factor will improve. If the motor is designed for power-factor correction in addition to its duty as a driving motor, then it will have an oversize field and exciter. By running the motor with more excitation than is necessary, it can be made to generate var and supply them to the system; this will help to improve the plant power factor.

7.17 Leading Power Factor

Any device having the characteristics of a capacitor will provide leading var. This does not necessarily mean that the system or load power factor will be leading. This will only occur when the leading var exceeds the lagging var in a system. It is very rare that an industrial plant operates with a leading power factor, and so we consider the term *leading power factor* to apply to equipment which will produce leading var.

Necessary calculations can then be made to determine the effect that the leading-power-factor device will have on the *system* power factor. All overexcited synchronous motors and underexcited generators—all capacitors and long lightly loaded power lines—will produce leading var.

7.18 Power-factor Correction

Here we are concerned with analyzing the *system* or plant load power factor. If it is anything less than unity (1), we must determine if it is economically feasible to correct it, and we must also establish how to correct it and to what level it can be economically improved.

The first step is to review the power-company rate schedule and examine the power-factor clause. This will establish the limits of economical power factor. Second, we establish all major items of equipment that provide leading and lagging var. Third, an examination of items of equipment is necessary, to see if there is one major item that is responsible for the poor or low power factor. Fourth, we establish if there are any synchronous motors that may be used for power-factor correction, and if there are, that they are in fact doing their job. When all these facts have been examined, it may then be established that some equipment must be purchased to correct the power factor.

7.19 Leading-var Generators

When considering power-factor correction, we are interested mainly in economics. Therefore, we are usually interested in how many leading var we can obtain for a given amount of money. We are also interested in equipment that will maintain the leading var in predictable amounts so that the corrective value will be maintained. A synchronous motor would be used only for a special condition; lightly loaded cables and lines (if they have a capacitive effect) are not constant in effect. This, then, leaves the capacitor. Here we have a choice of the series capacitor and the shunt capacitor. The former is inserted in the line and must carry the line current. It is a good regulator but an inefficient var generator. It is also a relatively expensive device. Alternatively the shunt capacitor is connected across the lines. It is relatively cheap and easy to install, and extra units can be added at a later date. Thus we can assume that the shunt capacitor is the routine equipment to use for power-factor correction, unless some special considerations exist.

7.20 Power-factor-correction Calculations

Power-factor-correction calculations are relatively simple. Any problem in making these calculations is usually due to lack of understanding of the subject. To ensure that it is fully understood, we will briefly point out that

1. Resistive loads do not use or generate var, and therefore they operate at unity power factor.
2. Magnetic fields contribute to a lagging power factor.
3. Capacitors, condensers, and any other equipment having a capacitive effect contribute to a leading power factor.

The unity-power-factor loads do not present any power-factor problems. Therefore, we can disregard all these types. The next consideration is whether the system has a lagging power factor, and if so, what it is. If the power factor is lagging, it will require leading var for correction. The amount can be calculated approximately. The word "approximate" is used in the sense that although the calculations are accurate, they are only as good as the information used. The power requirements of a system are dynamic and are never stable. Therefore, when this type of calculation is carried out, many assumptions are made. First, we will show that it requires approximately 1 kva to produce 1 hp in a medium-size motor, and around 1⅓ kva in the small fractional-size motors. The physics books state that 746 watts equals 1 hp, but this is output and does not indicate the efficiency and power factor of an ac motor. If we consider the power requirements for a 50-hp induction motor, we have

$$\text{kva} = \frac{50 \times 746 \times 10^{-3}}{0.85 \times 0.88}$$
$$= 50 \text{ kva input to drive a 50-hp motor}$$

where 0.85 is efficiency and 0.88 is the power factor. We must point out that the efficiency and power factor vary for different motors. The exact information is not usually available. In addition, the power factor of a motor changes with load; the more load the better the power factor.

For the purpose of our calculations, we will use the value of 1 kva per horsepower for motors over $7\frac{1}{2}$ hp.

When we make calculations for power-factor correction, we are simply calculating the number of var required by the load. Next, we simply add the equivalent amount of *leading* var (using capacitors) to cancel out the *lagging* var. This will then result in unity power factor. Very simple. To differentiate between var required (lagging power factor) and var supplied (from capacitors), we will use

kvar = var required = power factor lagging
ckvar = var supplied by capacitors (leading)

We will also introduce some simple equations for these calculations and avoid the right-triangle diagram. From the original right-triangle relationship it is possible to derive the following:

$$\cos \theta = \text{PF} = \frac{\text{kw}}{\text{kva}}$$

$$\tan \theta = \frac{\text{kvar}}{\text{kw}}$$

$$\sin \theta = \frac{\text{kvar}}{\text{kva}}$$

Now, we must consider what are the most likely problems requiring power-factor correction. There are two basic problems:

1. Plant power-factor correction for economical reasons
2. Release of capacity in cables and generators

7.21 Plant Power-factor Correction

What information is usually available to indicate a poor power factor in plants in the first place? There will probably be a power-factor meter, voltmeter, and ammeter. There may or may not be a wattmeter, or, if there is, there may not be a power-factor meter. Usually there are enough instruments to determine the power factor. This means that we can set up the power-factor equation. This also means that we now know the kva, kw, and $\cos \theta$; knowing the latter means that we can determine θ in degrees, as an angle. From this, we can determine $\sin \theta$ and tangent θ if required.

The kw or real power is the actual power required by the plant to operate, and usually this is reasonably stable. The power factor may change, causing a change in the kva required. We will assume then that the kw is usually the constant, and we will take the PF and kva as the variables. The equation using kw and kvar is

$$\tan \theta = \frac{\text{kvar}}{\text{kw}}$$

By using cross multiplication, we derive the following:

$$\text{kvar} = \text{kw} \times \tan \theta$$

If we have a particular load of 1,000 kw at 0.87 PF, we know that the $\cos 30°$ is 0.866 or 0.87, and from Table A.11, $\tan 30°$ is 0.577. Now, we can substitute these values in the equation:

$$\text{kvar} = 1{,}000 \times 0.58$$
$$= 580$$

This, then, is the kvar required by the load or plant system. To bring it to power factor of unity, we simply add 580 ckvar. Assuming we only want to raise it to 0.95 PF, how do we do this? This is quite simple. We calculate the power factor we require and subtract one from the other. Instead of doing two calculations, we can do just one.

$$\text{ckvar} = \text{kw} \times (\tan \theta_1 - \tan \theta_2)$$

where θ_1 is the original phase angle and θ_2 is the new or required phase angle. If 0.95 is equal to $\cos \theta$, then by checking the tables, we see that 19° is the angle and the tangent is 0.344. Now, by substituting into the equation, we have

$$\text{ckvar} = 1{,}000(0.58 - 0.34)$$
$$= 1{,}000 \times 0.24$$
$$= 240$$

We need only 240 ckvar to correct to 0.95 PF, and yet we require 580 ckvar to correct to unity, more than twice as much. Thus, we see that there is an economical factor in how much the extra capacitors would cost to raise the power factor from 0.95 to unity, and in how much of a saving there would be in the power-company billing. Obviously, there must be a number of trial calculations or a curve developed to find this economical point. Also, the management or budgeting department is usually the governing factor, and the money available determines how much correction can be made. It should also be obvious that if we are given a specific amount of ckvar, then the new power factor can be determined by manipulating our previous equation. For example, if we were only allowed 200 ckvar, what would the new power factor be?

$$\text{ckvar} = \text{kw} (\tan \theta_1 - \tan \theta_2)$$
$$= \text{kw} \times \tan \theta_1 - \text{kw} \times \tan \theta_2$$
$$\text{ckvar} - \text{kw} \times \tan \theta_1 = -\text{kw} \times \tan \theta_2$$

$$\frac{\text{ckvar} - \text{kw} \times \tan \theta_1}{\text{kw}} = -\tan \theta_2$$

$$\frac{-\text{ckvar}}{\text{kw}} + \tan \theta_1 = \tan \theta_2$$

By multiplying both sides by -1, we obtain an angle that would be in the first quadrant, that is, 0° to 90°. Now, if we substitute the values in the last equation, we obtain

$$\frac{-200}{1,000} + 0.58 = \tan \theta_2$$
$$-0.20 + 0.58 = 0.38$$

NOTE: *Recall that $\tan \theta_1$ is the original phase angle of 30° and $\tan 30° = 0.58$.*

The tangent of 0.38 has an angle of 21°, and cos 21° is 0.93. Therefore, the new power factor would be 0.93 or 93%.

We have used two simple equations; to recap, these are

$$\text{ckvar} = \text{kw} (\tan \theta_1 - \tan \theta_2)$$
$$\frac{-\text{ckvar}}{\text{kw}} + \tan \theta_1 = \tan \theta_2$$

This is all that is required for usual plant power-factor correction. Simply read the meters and substitute the values to determine the capacitors required. The capacitors have a ckvar rating, and any standard power-capacitor catalog will give the necessary voltage, price, and rating to suit the specific application.

7.22 System Capacity Release

We are fully aware that any conductors supplying equipment requiring var, carry both the kw current and the var current, and that the total current is the vector sum of the two. If it were possible to supply the reactive current from another source, then the supply conductors could carry only kw current. This, essentially, can increase the capacity of the supply conductors. It is also apparent that the var must be supplied by the substation or source of supply. As the power factor deteriorates, the substation or service must supply more kva to maintain the same kw output. Eventually the condition arises where no more kva is available. To correct this problem the first thing to do is to consider installation of a new service and substation, and second, release of system capacity by power-factor correction. The first step is obvious, and the price of installation can be readily obtained and can be expressed as money/kva. The second alternative is to release the kva used as reactive power and use it for kw power. This can be done by installation of capacitors throughout the plant at specific locations. These capacitors can supply the reactive power at the load. This will improve the system power factor. This can be shown in the equation for kw power:

$$kw = \sqrt{(kva)^2 - (kvar)^2}$$

We can see that the more we reduce the kvar portion of the equation, the more kw will be available; and if we removed it completely, or made it zero, then all the kva would become kw power. An evaluation of the load will show where capacitors can be installed and also how many ckvar can be added at specific loads. The total installation cost of capacitors can also be expressed in money/ckvar. Comparison of the substation money/kva and capacitor money/ckvar will indicate which approach should be followed. Consider an example of an induction motor requiring 100 amp at a power factor of 80% (0.80). If the 100 amp is the kva current, then the kw current will be $100 \times 0.8 = 80$ amp. The kvar current will be $100 \times \sin \theta = 100 \times 0.60 = 60$ amp. Notice that to find the kvar current, we used $I \sin \theta$ instead of the following equation:

$$kvar\ current = \sqrt{100^2 - 80^2}$$
$$= 60\ amp$$

We obtain the same answer, but the calculations are less laborious using $\sin \theta$. Also, there is more risk of error with the Pythagorean theorem.

7.23 Switched Capacitors

We know that the induction motor requires kvar, and that a shunt capacitor will supply ckvar. If we attached a capacitor of the correct size at the terminals of the induction motor, then this would supply the kvar current of 60 amp directly to the motor at the terminals. This means that the current in the supply conductors would only be 80 amp, and the supply transformer or source would only supply 80 amp, instead of 100. This is a saving of 20%. The same principle applies when considering main substation capacity; if enough capacitors can be installed at the loads, or at a close proximity to load centers, then similar results can be obtained.

Overloaded conductors can be saved from damage by this method, if the loads are requiring kvar.

Generators operating at full load conditions are usually supplying kvar, as well as real kw power. Here, the problem becomes more acute. The generator may be the only source of power available if it is in a remote location. Even unlimited money may not be able to purchase more power within the time period allowed. Power-factor correction, in this case, may be the only solution. Here, capacitors would probably be installed on the distribution lines and switched on and off as required. This can be done manually or automatically.

7.24 System Analysis and Capacitor Location

A typical industrial plant will usually have its own substation, numerous medium size induction motors, some gaseous tube-type lighting, such as fluorescent or mercury vapor, some incandescent lighting and resistive-type heaters, and probably one or a small number of synchronous motors. A study of the single-line diagram will show the different feeders supplying the various areas. The next approach is to determine the feeders supplying loads of poor power factor. A further analysis of loads will indicate which loads can have capacitors installed at the terminals of the load.

Depending on the degree of assessment required, the study can be continued until all individual branch circuits are considered. There is no specific rule that applies to all conditions for analysis, and there is no specific rule as to location of capacitors for all applications. Each problem must be considered individually with its attendant degree of accuracy necessary. There has been a great amount of information published by manufacturers on the subject of specific applications of capacitors. It is important that capacitors not be installed indiscriminately; all the possible effects that can arise from the installations should be taken into account.

7.25 Power-factor Calculations of Mixed Loads

Let us present a typical problem facing a plant engineer. Consider the single line in Fig. 7.2 and assume that you are asked to evaluate

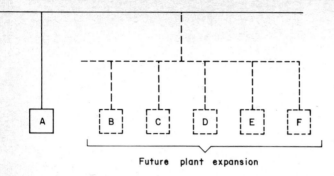

Fig. 7.2 Future plant expansion loads.

A. 2000-kva 87% PF/Existing plant load
B. 50-kva incandescent lights
C. 50-kva fluorescent lights
D. 50-hp fractional motors
E. 200-hp medium induction motors
F. 75-hp sync motor 0.8 PF

the effect of a future addition on the existing plant power factor. The single line may be the only information you are supplied with. If the following approach is used, the information is adequate, and fractional-motor kva is assumed to be 1.34 × hp. The synchronous motor requires 1 kva per hp and is always considered to have leading power factor.

PROBLEM: *What effect will new expansion have on existing plant power factor?*

Load	hp	kva	PF	sin θ	Lagging var	Leading var
Incandescent light.........	...	50	1	0.0	0	0
Fluorescent light..........	...	50	0.80	0.60	30	0
Fractional motors.........	50	65	0.71	0.71	46	0
Medium motors...........	200	200	0.85	0.53	106	0
Synchronous motors......	75	75	0.80	0.60	0	45
Existing plant............	...	2,000	0.87	0.5	1,000	0
kva total.................		2,440	kvar total........		1,182	45
			Subtract ckvar....		45	
					1,137	

$$\sin \theta = \frac{kvar}{kva} = \frac{1{,}137}{2{,}440} = 0.47$$

$$\sin^{-1} 0.47 = 28° \quad \text{and} \quad \cos 28° = 0.88$$

The new power factor would be 88% and would be slightly better than the old power factor. It is marginal whether it would be economical to spend money to improve this.

It may not be obvious, but if we consider the single line for the plant extension, we find that we have only one synchronous motor. It may not be running all the time the plant is in operation, and so we must consider what happens when it is shut down. Therefore

$$\sin \theta = \frac{\text{kvar}}{\text{kva}} = \frac{1{,}182}{2{,}440} = 0.485$$

$$\sin^{-1} 0.485 = 29°$$
$$\cos 29° = 0.87$$

Evidently it does not make any significant difference. Now, assume we wish to raise the new plant power factor to approximately 95%. How many ckvar do we require?

$$\begin{aligned}
\text{ckvar} &= \text{kw}(\tan \theta_1 - \tan \theta_2) \\
&= \text{kva} \times \cos \theta (\tan \theta_1 - \tan \theta_2) \\
&= 2{,}440 \times 0.88 (\tan \theta_1 - \tan \theta_2) \\
&= 2{,}147 (0.55 - 0.33) \\
&= 2{,}147 \times 0.22 \\
&= 471
\end{aligned}$$

$\tan 29° = 0.55$, $\cos^{-1} 0.95 = 18°$, and $\tan 18° = 0.33$ for $\tan \theta_2$.

This may not be economical if the plant was already in existence; however, the plant extension is not yet installed. Therefore, buying capacitors and installing them during normal construction is not as costly as a special project involving capacitor installation only. The accounting department will consider the capacitor installation as a fraction of a percent of total project cost. If it was a special project, it would be considered as 100% cost and would get much more scrutiny. When evaluating power-factor-correction work, remember that the department that supplies the money does not necessarily understand the engineering problems. Unless it can be presented to these individuals in terms they fully understand (usually money), it may not be approved, even though (from the engineer's point of view) it is absolutely necessary.

7.26 Capacitors on Distribution Lines

If capacitors are installed on distribution lines for power-factor-correction purposes, care must be taken to ensure that the voltage regulation[1]

---

[1] Perfect regulation would be unvarying voltage under all load conditions. Anything other than this would be termed "poor" or "acceptable" regulation and expressed in percent variation from standard.

of the line is not significantly affected. Normally, shunt capacitors are installed on distribution lines to assist in maintaining the line voltage between predetermined limits. Without going into the problem of regulation, we will show that capacitors will raise the voltage on a distribution line by a predictable amount.

$$\% \text{ volts rise} = \frac{\text{ckvar} \times d \times X}{(\text{kv})^2 \times 10}$$

where ckvar is three-phase capacitor rating, d is distance in units, X is line reactance in ohms/unit length, and kv is line-to-line voltage in kilovolts. For single-phase installation, the ckvar would be the single-phase capacitor bank, and the reactance X would be multiplied by 2 to give $2X$. The voltage rise would appear at the capacitor bank, and will remain the same amount during light and heavy loads. By means of automatic switching, the capacitors can be connected during heavy loading periods and disconnected during light loading periods.

7.27 Capacitors Installed at Transformers

When capacitors are installed at the transformer of a plant substation, a voltage rise throughout the plant distribution system can occur. The amount will vary with the distribution voltage, and also with the resistance to the reactance ratio of the system. The voltage rise is not usually high enough to cause high-voltage problems, and the following equation will give an adequate guide for evaluation:

$$\% \text{ volts rise} = \frac{\text{ckvar} \times \% X_t}{(\text{kva})_t}$$

where ckvar is the total capacitor rating, $\% X_t$ is the percent reactance or impedance of the transformer, and $(\text{kva})_t$ is the transformer kva rating. If we considered our previous problem of the plant extension, assume a 2,000-kva main transformer and a capacitor bank of 360 ckvar as a standard 460-volt bank. Assume that the transformer reactance is 5%.

$$\% \text{ volts rise} = \frac{360 \times 5}{2,000} = 0.9\%$$

This is less than 1%; therefore, further consideration is not necessary.

7.28 Motor and Switched Capacitor Installation

The induction motor is one source of low power factor, especially when it is lightly loaded. One approach to correcting this is to install a capacitor on the motor circuit. If it is installed on the load side of the motor circuit, then it will only be in operation when the motor is running. There are three optional locations for the capacitor. One is at the terminals of the motor, the second is at the load side of the contactor and ahead of the overload relay, and the third is on the line side of the motor starter. We will consider only the first one. The routine of selecting capacitors and accessories can usually be avoided by contacting the manufacturer first. If he is provided with the horsepower rating, speed, and type, he will advise on standard packages available to meet the requirements of the particular motor.

The National Electrical Code, Article 460, deals with capacitors. NEC 460-7 indicates that the kvar rating of the capacitor, connected on the load side of the motor, shall not cause the motor to have a leading power factor at no-load. This is easily calculated by the use of our previous equations. Considering these requirements, the current from the capacitors should not exceed the current of the motor when it is running at no-load. This can be checked by use of a clamp-on ammeter on the motor conductors when it is running at no-load. The current will be slightly higher than the magnetizing current, but adequate for this purpose. NEC 460-8(a) treats conductor ratings. The conductors connecting the capacitors with the motor *shall be not less than one-third the ampacity of the motor conductors* themselves, and they *shall not be less than 135% of the rated capacitor current*. The conductors used for the *capacitor circuits* shall not be less than 135% of the capacitor current rating.

NEC 460-8(b) considers overcurrent protection: (1) An overcurrent device shall be provided in each ungrounded conductor, *except* a separate overcurrent device is not required on the load side of a motor-running overcurrent device. The exception will apply to the installation at the motor terminals. This type of installation requires that the *motor-running overcurrent* protection be reduced. The motor conductors carry reactive current as well as kw current. Therefore, when a capacitor is installed at the terminals of a motor, the conductor no longer carries all the reactive current. By using our previous equations, we can calculate what the new current would be and size the overcurrent protection accordingly. See NEC 430-32, but apply the new lower value.

NEC 460-8(c) covers disconnecting means. All circuit breakers, contactors, and disconnecting devices, excepting safety switches (with or without fuses), should be rated at minimum 135% of rated capacitor current.

The safety-switch rating is covered by a NEMA standard for shunt capacitors which calls for a minimum of 165% rated capacitor current (fused or unfused).

NEC 460-10 insists that all capacitor cases shall be grounded in accordance with NEC 250. NEC 460-6 requires a draining of stored charge.

7.29 Calculation of Capacity

For some reason, it may be necessary to calculate the capacity in microfarads. Here then is the approach. First, we obtain the capacitor current. We will use 100 ckvar as an example:

$$I_c = \frac{100 \times 10^3 \text{ ckvar}}{480 \text{ volts}}$$
$$= 208 \text{ amp}$$

System voltage is equal to IZ, or current times impedance. Impedance is the vector sum of resistance and reactance. In this case, the resistance is negligible; therefore, we can state that $E = I_c X_c$, where X_c is capacitive reactance. By cross multiplying, we have

$$X_c = \frac{E}{I_c}$$
$$= \frac{480}{208}$$
$$= 2.3 \text{ ohms reactance}$$

Capacitive reactance is equal to

$$X_c = \frac{10^6}{2\pi f C}$$

By cross multiplying

$$C = \frac{10^6}{2\pi f X_c}$$
$$= \frac{10^6}{2 \times 3.14 \times 60 \times 2.3}$$
$$= 1{,}153 \text{ } \mu f$$

Capacitors in parallel add together; that is, 100 capacitors rated at 11.5 mf would give a total capacitance of 1,150 μf. For power work capacitors are rated in kva, and so this calculation is rarely needed.

Chapter Eight

System Study

8.1 Scope

A *system study* is an evaluation of the efficiency and effectiveness of an electric system. The system may be a high-voltage primary transmission or distribution system. Alternatively, it may be a secondary low-voltage distribution system. It may even be a combination of the high- and low-voltage systems. Before a system study is undertaken, the scope of the study must be outlined. The purpose and end objective of the study must also be specified *and specifically defined.*

8.2 The Single Line

The client will usually provide a single-line diagram outlining the system. First, this should be verified by a field trip (if possible) to ensure its accuracy. The single line is then revised to its "as built" condition. The next step is to define the beginning and end of the single line as it applies to the study. A new single line is made showing only the components and parts of the system involved. We now have a single line showing a starting point and finishing point as far as the study is concerned. Stay with this outline; it is easy to unintentionally drift outside the scope of the project.

8.3 Input Parameters

With the single line outlining the physical components, we now have to show the variables and limits of the input to the system. Any system must have a source of supply. This source may be indicated in the study, or it may not. In either case, we are only interested in a few items. We are interested in supply voltage. This will fluctuate. If possible, try to obtain a recording voltmeter and determine actual highest and lowest levels. The next important point is how much short-circuit kva is available. This will be supplied by the personnel operating the supply system, or it can be assumed (with a "Hold") if it is not readily available. The normal kva capacity of the system should also be known. The actual 24-hr loading should be obtained, if possible, on a recording ammeter, located at the input to the system. Measurement of power factor is also desirable when possible. This may be obtained by power-factor meter or wattmeter.

8.4 Reason for Study

At this point, it is assumed that all the necessary information is on the single line, i.e., voltages, equipment sizes, breaker and trip ratings, cable sizes, relays, etc. We now have the tools available. The next question is, What are we going to do?

System studies are made for any number of reasons:

1. Plant expansion
2. Poor voltage spread
3. Regulation problems
4. Overload coordination
5. System overloading
6. Increased short-circuit capacity
7. Change of voltage
8. Low power factor
9. Intersystem tie
10. Recloser installation
11. Relaying and metering

And so the list goes on. The main point to consider is the title. Write down the reason for the study as the title of the study, e.g., "System Study for Overload Coordination."

8.5 Study Layout

When a study is made, it is essentially a book describing the system, its problems, the reasons for these problems, and remedial action. The approach, then, is to lay out a study in the form of a book. First make up a preliminary table of contents. This will outline the various steps. Next, it must be decided which graphs, charts, and diagrams will be required. After this is completed, we should refer back to the contents. We now consider the first item. We list the actual subitems by task under the general headings. For example the main heading may be "Input Parameters." This would then be listed by task as follows; Recorded Voltages 24-hr—Record Amperes 24-hr—Short-circuit KVA Available—Supply Capacity, etc. This contents listing is the most important part of the study. The cost, time evaluation, manpower, instruments, etc., are all evaluated or estimated on the basis of this table of contents.

8.6 Degree of Complexity

We are all familiar with published engineering studies, with pages of calculus, haversines, triple integrals, etc. These are probably excellent papers for theoretical discussion. The problem is who can understand them? Another problem is that a paper or study can predict to the nth degree what should happen. Unfortunately, the machine operator making the equipment did not see the paper, and probably would not have understood it if he had, but nevertheless he carries on, turning out parts for equipment. He makes these parts with a limited degree of accuracy; therefore, all the technical papers in the world will not change this. All that these studies intend is to present a set of facts to a set of individuals who have a specific level of understanding. It is important that the study contents are limited in degree of sophistication to the degree of implementation intended. It is nice to be impressive, but a client pays for useful and accurate information, which is immediately and obviously factual.

8.7 Protective-device Coordination

Any system has a series of protective devices. These can be fuses, circuit breakers, and relays. Possibly a recloser and sectionalizer can be included in this group, although more probably they belong in a class by themselves as a system monitor and/or isolator. All these protective devices have different speeds of operation. The various types of relays are innumerable, but their basic function is to shut the system down if certain limits are exceeded. All these devices, then, are watching the system. When one device operates, we may decide that it was the wrong one. A small relay may shut down New York City when it should only shut down a line section without service interruption to New York. For coordination analysis the usual system study approach is followed. The main item required is *operating-characteristic time curves* for the various protective devices. By an evaluation of these device time curves, a reasonable prediction can be made of the sequence of operations.

8.8 Fuses versus Circuit Breakers

A main protective device usually falls into the category of a fuse or a circuit breaker. Each has its own advantage and disadvantage. In a three-phase system, when fuses are installed, the possibility exists that only one fuse may "blow." This leaves power on the other two phases, resulting in a "single-phase" condition. The normal three-phase induction motor requires three-phase power to start, but will continue running if the power to one phase is removed. Unfortunately, the motor output is reduced considerably. If the load requirement exceeds the reduced motor output, then the motor winding will overheat and possibly burn out. If a distribution-line fuse opens one phase, then this can affect a large number of motors. This is the obvious disadvantage of the fuse; also, spares are required. The advantages are low cost and fast operating speed. Fuses can be obtained with various characteristics, such as current limiting and surge override, and so they should not be ruled out.

8.9 The Circuit Breaker

This device is the generally accepted overload device. It is much more versatile than the fuse. It opens all three phases at the same time. It can be used as a load break disconnect. It can be controlled by external devices such as small relays. These can cause the circuit breaker to open on deviations in frequency, voltage, current, phase angle, temperature, and numerous other functions. We see, then, that its main advantage is versatility. The circuit breakers available begin with a small plastic case, which operate on thermal buildup. The range extends to large, highly sophisticated air-blast and vacuum-type power circuit breakers, controlled by a complex network of relays.

8.10 The Recloser

The recloser is basically a circuit breaker; it is designed to open on a fault, wait for a short time, and then reclose. It will repeat this procedure a predetermined number of times; if the fault is still present, it will lock out, deenergizing the faulty circuit.

8.11 The Sectionalizer

The sectionalizer is not a circuit breaker. It does not break the load current. Its function is to count the number of reclosures on the circuit breaker ahead of it. After a specified number of operations of the circuit breaker, the sectionalizer will open and stay open—isolating the fault and allowing the breaker to reclose.

8.12 The Evaluation

The fuse and circuit breaker each have a useful function, and should complement each other in a system. The fuse should be used as a backup safety device and should be rated high enough so that it will not "blow" under normal overload conditions. Its function is protection against the sudden massive short-circuit fault that the circuit breaker does not have time to interrupt. The circuit breaker, in contrast, should be used for controlling heavy overloads with a relatively low speed buildup.

8.13 Relays—Analog Type

Fortunately, or unfortunately, relays are so numerous in type and characteristics that the study of relaying is beyond the scope of this book. It is a specialized field, requiring separate study and a familiarity with system dynamics. A simple relay protective system can be designed with general knowledge, but any type of complex system should be checked by a relay expert.

8.14 Relays—Digital Binary Type

In contrast to the analog type of relay, which monitors variable quantities, we have the simple relay which is ON or OFF. This can be compared with the number system, known as the binary system. ON is 1 and OFF is 0. If we apply power to the relay, it will operate. It will then open or close one or more contacts, depending on whether they were normally open or normally closed to begin with. This type of relay is not normally associated with the analog type. The digital[1] type of relay is intended for design of automatic control systems and is relatively simple. It is only mentioned here to point out that it exists under the same name, although in actual fact under an entirely separate category. The two types, then, would be "protective relays" and "control relays."

[1] "Digital" is used in the sense of "digital computer." The digital device (computer) counts in numbers with a specific yes or no decision for each number, for example, 0 or 1 for binary. Alternatively the analog device constantly monitors dynamic signals; for example, a thermometer is an analog device.

Chapter Nine

The Short-circuit Study

9.1 Short Circuits

Short circuit is the term associated with a specific fault condition. If we very briefly review Ohm's law, we see that the amount of current I in a circuit is limited by the resistance R built into the circuit; that is, $I = E/R$.

If the circuit is rated for 20 amp and is actually loaded to 10 amp, with a supply voltage of 240 volts, the resistance in the circuit must be $240/10 = 24$ ohms.

If a jumper of 20-amp conductor (No. 12 AWG) approximately 2 in. long is placed across the terminals of the two-pole 20-amp circuit breaker, we now have a resistance of approximately 0.00027 ohm in parallel with a resistance of 24 ohms. Solving for I gives

$$\frac{1}{R_T} = \frac{1}{R_1} + \frac{1}{R_2} + \cdots + \frac{1}{R_n}$$

$$= \frac{1}{0.00027} + \frac{1}{24} = 3{,}703 + 0.42 = 3703.42$$

Neglect the 0.42 as insignificant.

$$R_T = \frac{1}{3{,}703} \quad \text{or} \quad 0.00027 \text{ ohm total resistance}$$

We see, then, that the load resistance is negligible when added to the resistance of the jumper. Now we evaluate the current in the circuit. With the new low resistance added, we find that we have

$$I = \frac{240}{0.00027} = 888{,}888 \text{ amp (or max available current)}$$

NOTE: *The 888,888 amp is theoretically what the current would be if that much current were available. However, it would be a virtually impossible value, and in practice 8,888 amp would be more realistic.*

This, then, is the effect of a short circuit.

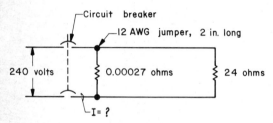

Fig. 9.1 Solving for I where $I = E/R_T$.

9.2 Breaker Interrupting Capacity (IC)

If the system supplying the circuit is capable of maintaining 8,888 amp, then the electromagnetic forces and gases within the circuit breaker will be extreme. If the breaker rating is less than 8,888 amp IC, then it may possibly disintegrate. Disintegration leads to unintentional ionization of the air in the breaker compartment. A continuous arc is then created. This is maintained until a protective device removes the power, or until the conductors blow themselves apart sufficiently to cause an open circuit. This condition can cause tremendous damage. This is the type of situation the designer should worry about when he thinks of a "short-circuit condition."

9.3 Bolted Short Circuit

When we actually consider a short-circuit study, we imagine the *jumper* as a large piece of copper bar bolted across the three-phase conductors. This is then considered as practically zero resistance, and is called a *bolted short circuit*.

9.4 Short-circuit Current

Thus we see that a short-circuit condition differs from the normal current condition only by virtue of accidental decrease in circuit impedance. It is not a different source of current that magically appears. It is only a load current that is very rapidly increased as far as the overcurrent device is concerned. When the increase exceeds the design limits of the equipment, then we encounter the destructive forces which make it necessary to carry out the short-circuit study.

9.5 The Short-circuit Study

The first step in conducting a short-circuit study is to draw the complete single-line diagram. The diagram will show all transformers, motors, conductors, circuit breakers, fuses, etc. All items should be identified by a number or letter. Circuit breakers and switches are classified as having zero impedance. Therefore, they need not be considered. On

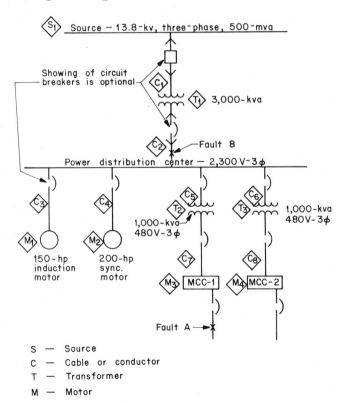

S — Source
C — Cable or conductor
T — Transformer
M — Motor

Fig. 9.2 Single-line identification of reactances.

the single-line diagram, we will select the points where we locate an imaginary fault.

9.6 Fault Location

The usual reason for working a short-circuit study is to determine whether a protective device will withstand the available short-circuit stresses. The situation is usually the evaluation of a new design, and the designer must select the necessary new equipment. Alternatively, it may be an existing plant that is being up-rated by addition of a new or enlarged power source. In this case, the existing circuit breakers should be proved by a short-circuit study, to show that they are adequate.

We are interested in the maximum short-circuit current that the circuit breaker will have to withstand. We assume that this will occur at the load terminals of the breaker. If we evaluate it at this point, then we know that a fault occurring at any point after that will be acceptable. That is until it reaches another breaker. Then the evaluation must be repeated.

9.7 Equipment Impedance

Impedance in an ac circuit is defined as the *total* opposition to a flow of current. We have three variable parameters to consider. These are opposition due to resistance, capacitance, and inductance. When a current is related in time to voltage, we find that

1. With resistance only, the current will be in phase.
2. With capacitance only, the current will lead the voltage.
3. With inductance only, the current will lag the voltage.
4. With resistance and capacitance and inductance, the current may be in phase, lead, or lag, depending on the values of the parameters

9.8 Impedance Parameters

Resistance is measured in ohms, capacitance is measured in farads, and inductance is measured in henries. The units in this form are not conducive to relative manipulation. To allow manipulation of all three quantities together, we assign an equivalent ohmic value to capacitance and inductance. We then call this *reactance*. To distinguish capacitance from inductance, we use the following:

$$X_c = \text{capacitive reactance}$$
$$= \text{ohms} = \frac{10^6}{(2\pi f)C}$$

where f = frequency and C = capacitance in μf.

$$X_L = \text{inductive reactance} = \text{ohms} = 2\pi f L$$

where L = inductance in henries.

Now we have each parameter rated in ohms. When we combine these quantities, it must be done vectorially. Impedance Z is the total opposition to a flow of ac current and is expressed in cartesian form as

$$Z = r + jX$$

From our mathematics section in Part 2, we know that the magnitude of the vector would be

$$Z = \sqrt{R^2 + X^2}$$
$$= \sqrt{R^2 + (X_L - X_c)^2}$$

Here, then, we now have the three basic parameters in the single equation. This now gives us the freedom to manipulate which we did not have before.

9.9 Percentage Values

To further simplify short-circuit calculations, we change all the actual ohmic values to percentage values. By use of a formula, we relate the actual reactance ohms to the voltage and kva ratio of the equipment.

$$\% \text{ ohms reactance} = \frac{(\text{ohms reactance})(\text{base kva})}{(\text{kv})^2 \times 10}$$

Manufacturers publish tables indicating the percent impedances, or percent reactances, for their equipment. Recalling that the difference between impedance and reactance is the resistance added vectorially, we can now further simplify by asking the following question: Is the short-circuit study accuracy affected by ignoring the resistance values and working with reactance only?

9.10 Degree of Accuracy

Let us reexamine the reason for contemplating a short-circuit study at all. We will also consider the degree of accuracy necessary.

The normal approach is to have a circuit breaker existing, or one already selected by virtue of lowest cost or preferred type. The item in question will have a specific short-circuit rating. It is now necessary for the record (job file) that a statement be made that this equipment is adequate to withstand the short-circuit capacity of the system. It is now up to the responsible individual to justify the statement with a formal calculation.

The system and equipment may be so simple and extreme in rating that a mental calculation can satisfy (even this should be recorded). Or an alternative would be where a large expensive circuit breaker exists and the available short-circuit capacity is increased. This can happen if a small town diesel electric system is replaced by a large utility system, with infinite bus capacity. Now, the economics warrants a very accurate study to determine that the breaker will handle the increased short-circuit current. A calculation can be made using reactances only. This will usually suffice to indicate that the equipment rating is adequate. If it is a borderline result, where the breaker is slightly under the required rating, then the calculation should be repeated, using all impedances, and including *all* items, such as circuit-breaker contacts, small cables, and all available reactances and resistances. This will usually suffice. If all results are exhausted, then we revert to the system study. We then determine how the circuit impedance can be increased to limit the short-circuit current to an adequate level. Consider, for instance, installation of a line inductance (reactor).

In line with our statements on degree of accuracy, we will deal only with reactance. Where impedance Z is given, we will assume that this is reactance. For instance, a transformer is rated in percent Z at 5.7%. In the calculations, this would become

$$\% X = 5.7 \quad \text{or} \quad \% \text{ reactance} = \text{approximately } 5.7\%$$

If we require further accuracy, we would, of course, redo the calculation with impedances.

9.11 Equipment Rating—Transformer

The next step after the single line is to list all the reactances of the various components. These are then converted to a common base. This follows the same principle when dealing with lengths ($\frac{1}{8}$ in. + $\frac{1}{4}$ in. + $\frac{1}{2}$ in. + $\frac{7}{8}$ in. + $\frac{7}{16}$ in. = ?). To manipulate these figures, they must be changed to a common base. If we choose sixteenths as the base, then all the numbers are changed to the equivalent number of sixteenths (that is, $2 + 4 + 8 + 14 + 7 = \frac{35}{16}$). The same principle applies to changing the reactances.

The preceding single line shows item T_1 as a 3,000-kva transformer, with a primary voltage of 13.8 kv and secondary voltage of 2,300 volts. Typically, this is virtually the only information available. Consulting a published table for power transformers, approximate impedances, we would possibly have the following information:

TABLE 9.1 Power Transformer Impedances

HV, kv	LV, kv	Self-cooled or water-cooled, %	Forced-oil-cooled, %
15 or lower	15	5.5	6.75
25	15	5.5	8.25
32.5	15	6.0	9.0

We would have to assume the type of cooling if it was not known. We select the oil type at 6.75 Z at 3,000 kva.

9.12 Base KVA

If we arbitrarily select a 10,000-kva base, then we must replace the impedance with its equivalent at 10,000 kva, instead of the 3,000 kva, as rated. We use the following equation:

$$\frac{(\text{Base kva})(\% \text{ reactance})}{\text{Equipment kva}}$$

For the transformer, we would have

$$\frac{10{,}000 \times 6.75}{3{,}000} = 22.5\%$$

If we had selected a 1,000-kva base, we would have

$$\frac{1{,}000 \times 6.75}{3{,}000} = 2.25\%$$

The selection of the base depends on personal preference, either using large numbers, as shown in the 10,000-kva base, or small numbers, as indicated by the 1,000-kva base.

9.13 Rotating Equipment

When considering motors, we have a situation where, during a fault, the voltage at the motor drops considerably. The motor will then begin to slow down. Due to the inertia, it will rotate for a short period of time. It takes time for a magnetic field to collapse. Therefore, we momentarily have the condition where the motor acts like a generator. Even if the breaker opens fast, the fault will still be supplied with power

TABLE 9.2 Condensed Table of Multiplying Factors and Rotating-machine Reactances
To Be Used for Calculating Short-circuit Currents for Circuit-breaker, Fuse, and Motor-starter Applications

Classification	Circuit voltage	Location in system	Multiplying factor	Rotating machine reactances to use		
				Generators, synchronous converters, synchronous condensers, frequency changers	Synchronous motors	Induction motors
Power Circuit Breakers							
				Interrupting duty		
Eight cycle or slower (general case)........	Above 600 volts	Any place where symmetrical short-circuit kva is less than 500 mva	1.0	Subtransient	Transient	Neglect
Five cycle........	Above 600 volts		1.1	Subtransient	Transient	Neglect
				Momentary duty		
General case........	Above 600 volts	Near generating station	1.6	Subtransient	Subtransient	Subtransient
Less than 5 kv........	601 to 5,000 volts	Remote from generating station (X/R ratio less than 10)	1.5	Subtransient	Subtransient	Subtransient
High-voltage Fuses							
				Three-phase kva interrupting duty		
All types, including all current-limiting fuses........	Above 600 volts	Anywhere in system	1.0	Subtransient	Transient	Neglect
				Maximum rms ampere interrupting duty		
All types, including all current-limiting fuses........	Above 600 volts	Anywhere in system	1.6	Subtransient	Subtransient	Subtransient
Non-current-limiting types only........	601 to 15,000 volts	Remote from generating station (X/R ratio less than 4)	1.2	Subtransient	Subtransient	Subtransient

High-voltage Fused Motor Starters

			Three-phase kva interrupting duty		
			Subtransient	Transient	Neglect
All horsepower ratings	2,400 and 4,160 wye, volts	Anywhere in system	1.0	Subtransient	Neglect
			Maximum rms ampere interrupting duty		
			Subtransient	Transient	Subtransient
All horsepower ratings	2,400 and 4,160 wye, volts	Anywhere in system	1.6	Subtransient	Subtransient

High-voltage Motor Starters

			Interrupting duty		
			Subtransient	Transient	Neglect
Circuit breaker or contactor type	601 to 5,000 volts	Anywhere in system	1.0	Subtransient	Neglect
			Momentary duty		
Circuit breaker or contactor type	601 to 5,000 volts	Anywhere in system	1.6	Subtransient	Subtransient
Circuit breaker or contactor type	601 to 5,000 volts	Remote from generating station (X/R ratio less than 10)	1.5	Subtransient	Subtransient

Apparatus, 600 Volts and Below

			Interrupting or momentary duty		
Air circuit breakers or breaker-contactor combination motor starters	600 volts and below	Anywhere in system	1.25	Subtransient	Subtransient
Low-voltage fuses or fused combination motor starters	600 volts and below	Anywhere in system	1.25	Subtransient	Subtransient

from the motors. This is known as *short-circuit contribution*. This applies to induction motors and generators. The circuit dynamics involved are complex, and so we will not try to explain the theory.

9.14 Equipment Reactances

There are three reactances involved in rotating equipment:

1. Subtransient reactance $X''d$
2. Transient reactance $X'd$
3. Synchronous reactance Xd

Subtransient reactance $x''d$ appears at the *instant* the short circuit occurs. It controls the current flow during the initial few cycles after the fault.

Probably the reactances of motors will not be known. Therefore, another assumption must be made. As $X''d$ will vary, a good general figure would be 25% for motors of less than 600 volts and 20% for large motors of over 600 volts. For known special motors, with unusual starting characteristics, compare with the following formula:

$$\% \ X''d = \frac{100}{\text{stalled rotor current}^{1}} \quad \text{at rated kva base}$$

Transient reactance $X'd$ appears effective following the subtransient period, and can last approximately ½ sec or longer.

Whether subtransient or transient reactances are used depends on the type of protective device and the speed of operation.[2]

Synchronous reactance Xd is not effective for several seconds after the short circuit, and will not be used for protective-device studies.

9.15 Momentary Rating

A device for interrupting a short circuit must be able to withstand the *initial* shock stresses developed at the *instant* the short circuit occurs. The device must then contain these stresses, until the device opens the circuit. This is known as the *momentary rating*.

9.16 Interrupting Rating

During the time that the momentary rating is in effect, the equipment stresses are high but decaying slightly. As soon as the equipment can

---

[1] Where "stalled rotor current" is in multiples of normal current, that is, "5 times normal."

[2] See Table 9.2.

operate, it will open the circuit. To do this, it must be able to handle the momentary stresses existing at that time without damage. This, then, is termed the *interrupting rating*.

9.17 Impedance Diagram

The short-circuit study is relatively simple, once the impedance diagram is established. The important step, then, is setting up the impedance diagram.[1] We begin with the single line, as previously mentioned. Next we draw a similar single line, replacing all equipment symbols with a resistor symbol. The impedances will then eventually be reduced to a single impedance. This serves as the master plan for the short-circuit study. We must now decide where to locate the fault, and whether we require interrupting or momentary rating. We combine reactances with the same rules as resistors.

Series resistors $\quad R_T = R_1 + R_2 + R_3 + \cdots + R_n$

Parallel resistors $\quad \dfrac{1}{R_T} = \dfrac{1}{R_1} + \dfrac{1}{R_2} + \dfrac{1}{R_3} + \cdots + \dfrac{1}{R_n}$

The approach to laying out the impedance diagram is relatively simple. It is necessary to remember some basic rules:

1. Show all sources of fault current, i.e., induction motors, generators, and synchronous motors.
2. Replace all components having resistance and/or reactance with a resistor symbol.
3. Identify these components with letters.
4. Show all power-transformer secondaries feeding an induction-motor load whether motors are indicated or not (unless all load is accounted for).
5. Join all components by an "infinite bus."
6. The *source* is not the "infinite bus,"[2] but is shown as a separate reactance.
7. Rearrange reactances carefully into parallel and series groups.
8. When considering momentary rating, include all induction motors and use subtransient $X''d$ reactances.
9. When considering interrupting rating, neglect all branches feeding pure induction motors and use only transient reactance $X'd$ except below 600 volts.

[1] Even though we are using reactances, the accepted term is *impedance diagram*.
[2] The "infinite bus" can be considered a conductor connecting all the machine's neutral points together.

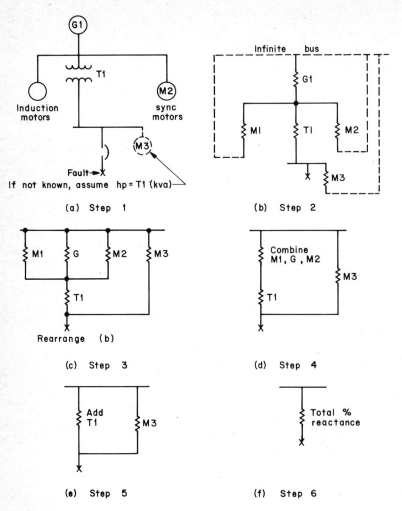

Fig. 9.3 Impedance diagram, Example 1. Neglect cables for example only. Calculation is for momentary rating.

We notice in Example 1, Step 2, that we have reactances connected at only one end. The general procedure to follow here is to connect the unconnected ends to the source (infinite) bus, except for the reactance(s) terminating in the fault.

Now we will consider a more involved system in the Example 2, Fig. 9.4. Notice the two following points: The motor loads $m1$, $m2$,

The Short-circuit Study

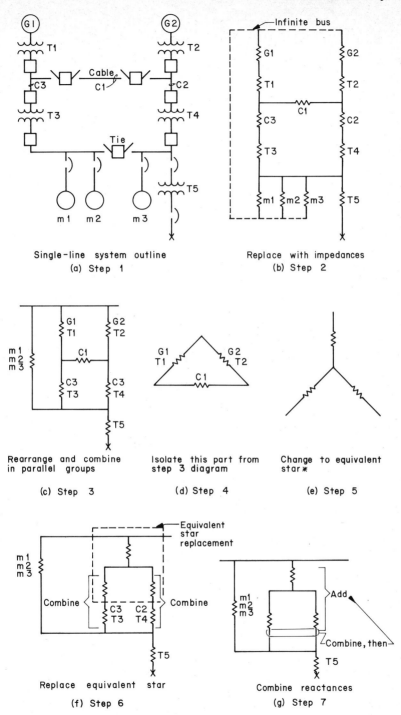

Fig. 9.4 Impedance diagram, Example 2; example with tie breaker closed.

118   Electrical Design Information

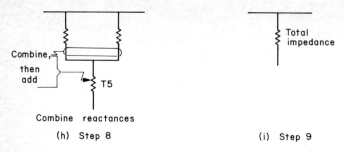

(h) Step 8   (i) Step 9

* See sec. 9.18 for star-delta conversion

Fig. 9.4 (continued) Impedance diagram, Example 2; example with tie breaker closed.

$m3$ in Step 2 were connected to the "infinite" bus. The second important point was the star/delta conversion. It should be recognizable that the reactances $G1$, $T1 - C1$, $- T2$, $G2$ form a closed delta in Steps 2 and 3. By changing this to an equivalent star connection, we have a simple series parallel equivalent.

9.18 Star/Delta Conversion

The star/delta conversion is accomplished by the use of the following formulas:

$$B = \frac{ab + ac + bc}{b} \qquad b = \frac{CA}{A + B + C}$$

$$C = \frac{ab + ac + bc}{c} \qquad c = \frac{AB}{A + B + C}$$

$$A = \frac{ab + ac + bc}{a} \qquad a = \frac{BC}{A + B + C}$$

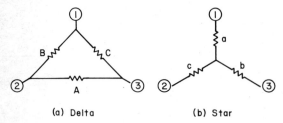

(a) Delta   (b) Star

Fig. 9.5 Star/delta conversion numbering method.

We see then that if we assign the numbers 1, 2, and 3 to the corners of a delta, we can solve for a, b, and c. This produces the star connection with corresponding 1, 2, and 3 connections.

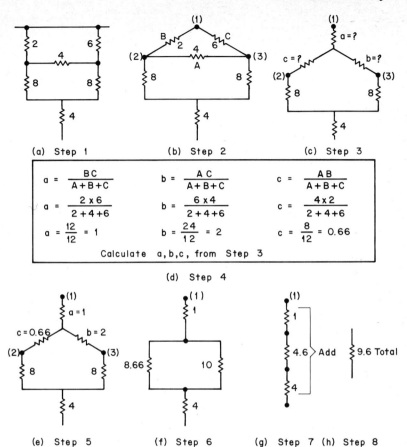

Fig. 9.6 Star/delta conversion application, Example 3.

9.19 Short-circuit Magnitude

Once the impedance diagram is set up, it is a simple matter to reduce to a single reactance. When this is obtained, we plug the value into a simple equation:

$$\frac{100}{\% X} \text{ (base kva)} = \text{short-circuit kva}$$

To obtain the short-circuit current, we use the general formula for three-phase current:

$$I = \frac{\text{kva}}{\sqrt{3} \text{ kv}} = \frac{\text{kva} \times 1{,}000}{\sqrt{3} \text{ volts}}$$

If we require the current in the first step, we simply combine the two equations.

$$I_{sc} = \frac{100}{\% X} \frac{\text{base kva}}{\sqrt{3}\,\text{kv}}$$

This will give the symmetrical short-circuit current.

9.20 Asymmetrical Current

The theory of asymmetrical currents is complex. Therefore, we will try to avoid it. Briefly, whether a sine wave of current is symmetrical about the zero axis depends on two things: (1) the X/R ratio, i.e., reactance to resistance, and (2) the exact time that the fault occurs. There is only one exact time and X/R ratio that will produce symmetrical current. The chance of this occurring is remote. Therefore, we must consider the worse condition, i.e., the asymmetrical current.

Considering the degree of accuracy again, we will refer to the multiplying factors given in Table 9.2. By multiplying the symmetrical short-circuit current, or short-circuit kva, by this number, we obtain the asymmetrical rating. For example, with hypothetical figures, at 4,160 volts:

$$I_{sc} = \frac{100}{50\%} \frac{10,000}{1.73 \times 4.16} = 2,779 \text{ amp symmetrical}$$
$$= 2,779 \times 1.5 = 4,168 \text{ amp asymmetrical}$$

for less than 5 kv, remote from the generating station, and X/R less than 10.

9.21 Rules of Thumb

In actual practice, in any trade, we encounter empirical values. These are values obtained by experience and observation, rather than theoretical calculations. The values, once developed, are sometimes used as a basis for theoretical calculations and are surprisingly reliable, even if not extremely accurate. Consider, for instance, the transformer. A transformer is designed to withstand certain stresses produced by a short circuit. The designer uses a multiple of approximately 20 to 25 times full load current. He then designs the transformer to limit any short-circuit current to this value. If this is the case it must follow that any fault current will be limited to at least 20 times full load current of the transformer ahead of the fault.

This is not a guaranteed answer, but if the short-circuit calculations are far more excessive than this, then a review of the calculations and parameters should be made.

9.22 Motor Control Center

Some manufacturers claim that any low-voltage motor control center will not pass more than 25,000 amp, regardless of the amount of short-circuit current available. This was derived by repeated tests. Once again, this is not a specific answer, but anything higher than 25,000 amp at an MCC would warrant a review of the calculations and further analysis.

9.23 Summary

It is obvious to the knowledgeable that this has been a very brief review of short-circuit calculations. To do justice to the subject warrants a whole book which would include the application of symmetrical components on unbalanced systems. The numerous tables of reactances have been purposely omitted for general items. These can be obtained in any general catalog or reference book. What we have attempted to do is to provide a memory jogger for the "rusty," and a "brake" for the "overenthusiastic" graduate whose employer does not have the many hours available for an exhaustive study that is probably not required anyway.

Chapter Ten

Instrumentation and Control Circuits

10.1 Project Conception

In most cases, a project will be outlined by the chemical or mechanical departments. Any project deals with input of raw materials and output of the end product. The research engineer conceives the process and the development engineer designs a system that is commercially feasible for production. To create a workable system, the development engineer "makes" a set of instructions. These are usually in the form of a set of specifications. These specifications will state exactly what the raw materials will be and each successive "change of state" they should go through. He will advise on necessary alarms required, safety shutdowns, and all other control items.

10.2 Flowsheet

When a project is turned over to the consulting engineer for design of construction drawings, the first act of the consultant is to interpret the client's specifications. To do this, a flowsheet is provided or developed. This is a drawing which shows the raw materials passing through each piece of equipment; it shows motors, heat exchangers, filters, tumblers, blenders, or any other pieces of equipment necessary.

Also, all instrumentation and electrical interlocks are usually shown. A good complete flowsheet with a set of specifications should provide all the information necessary for the control-circuit design.

10.3 P&ID

A P&ID sheet is the same thing as a flowsheet. "P&ID" is an abbreviation for process and instrumentation diagram. All the information in Sec. 10.2 is applicable to the P&ID as well as to the flowsheet.

10.4 Analog Instruments

Instrumentation takes many forms, but as far as the electrical designer is concerned, there are only two groups. The first group consists of instrumentation in the true sense, i.e., analog measuring devices. These instruments usually require only a power supply. In some cases, the wiring has to be provided for the signals into the instruments. If this is the case, the manufacturer will provide the drawing indicating the wiring necessary.

10.5 Digital Instrumentation

In the second group of instrumentation we have devices such as flow switches, pressure switches, temperature cutout devices, and numerous other types of devices. All these devices are really misnamed as instrumentation. They are actually *control devices*. The term *digital* refers to the ON or OFF function of the switch. This, in effect, is a binary unit like the relay. In this case, the actual operation of the switch is caused by an external force, rather than a solenoid as in the case with a relay. The control devices, then, are the main concern of the design engineer.

10.6 Protective Relays

We have previously mentioned relays. There are two basic types. The *protective relay* is essentially an analog device with a digital readout. The relay itself monitors the dynamic functions, but the readout is in the form of a contact, which closes or opens at a preset limit. This contact can then be used to initiate one or more other functions. The sensitivity of these relays is adjustable by changing the time/distance relationship of the two contact portions. The relays are usually provided with an instruction book, which gives the time/trip curves for that particular relay.

10.7 Control Relays

A control relay is a very simple device in contrast to the protective relay. It is not a system sensing device and it is not capable of monitoring dynamic system functions. The basic construction of a control relay consists of an electromagnet operating on a plunger. When the power is applied to the coil, it operates. When the power is removed, it returns to its previous position. It is hard to visualize anything simpler. By attaching contacts to the moving part and stationary part of the relay, we now have a device which can complete a circuit or open a circuit. We have then a yes or no indicator (0 or 1) in binary form.

10.8 Logic Arrangement

The design engineer is required to produce a control system that is automatic or semiautomatic. This control system must conform to the flowsheet (P&ID) and also the specifications. By arranging all the device contacts in a particular logic sequence, the automated (or semi-automated) system is achieved. To recap, the devices available are

1. Control devices (digital instrumentation)
2. Protective relays
3. Control relays

Each item provides a digital signal in the form of a contact.

10.9 Control Contacts

All the previous devices are at times "on the shelf" in storage. When considering a device, always present it in an "off the shelf" state. A device may have one or more contacts. These may be normally open (N/O) or normally closed (N/C). When we say "normally," we are

Fig. 10.1 (a) Normally open contact; (b) normally closed contact.

referring to the "on the shelf" state. There are various symbols for showing contacts, and, as "standards" vary, we will use only the two symbols shown in Fig. 10.1 for all contacts. If the contact is a pressure switch, it will be noted PS and the abbreviations explained in the notes, or it will be represented as in Fig. 10.2.

10.10 Relay Variations

Control relays have slight modifications "built in." These modifications are very limited in scope and usually consist of time-delay functions, latch-in, or overlapping contacts. A time-delay relay may vary in construction, depending on the time duration. The construction of the

Fig. 10.2 (*a*) Normally open pressure switch; (*b*) normally closed pressure switch.

device is of no concern to the design engineer, as long as it meets his requirements. The same applies to other variations. The basic approach, then, is to ignore the actual device and concentrate on the logic sequence. After the circuit logic is complete, then a selection of devices from the catalogs becomes routine. If a device cannot be purchased economically, then review the logic and modify the circuit to use a readily obtainable device.

10.11 Selector Switches

Switches used in control systems may be simple or complex. The HAND/OFF/AUTO switch is a simple three-position switch. The multicontact multiposition, hand- or motor-operated selector switch can become complicated. Variations of selector switches are numerous and virtually unlimited in availability if money is no consideration. The only real problem, however, even with the very complex switches, is to be able to keep track of the operating sequence. This is usually controlled by a diagram plotting contact closures against position. We see in Fig. 10.3 that in position 3, contacts 1, 2, 6, 7, 11, and 12 are marked with an X. This indicates that in position 3, all these contacts

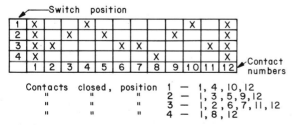

Fig. 10.3 Twelve-contact four-position selector switch.

will be closed. All contacts not marked are open. This type of diagram allows us to express a switch operation very simply.

10.12 Limit Switches

Another very common control device is the limit switch. This device is very simple and consists of a spring-loaded switch (or switches) in a box with an extended lever, which operates the switch. The variations are usually in the operating arm. A conveyor would usually use limit switches; as the product moves down the conveyor, it would apply pressure to the lever and operate the switch and return to normal position after transition.

A variation on this is the photoelectric cell. A beam of light is directed into a receiver containing "light-sensitive" material. When the light beam is broken, the receiver sensor causes a relay to change its state. The light beam may be visible, or, by adding an infrared filter, it can be made invisible.

10.13 Control-circuit Parameters

The simple devices we have mentioned are basically all that is necessary to design an automated control system for normal industrial use. The most sophisticated high-speed digital computers cannot do any more; they can only do it faster. With high-speed computer switching the transistor is required, and with it, the necessary environmental control. In industry, extreme high-speed switching is not essential. But reliability and simplicity are. We may then consider the control circuit to be similar to a very slow computer.

The flowsheet and specifications present the instructions for the sequence of operations. This, then, is the basis for the *program*. "Program" is a word that has evolved with the computer age and means, very simply, *predictable operation.*

10.14 Program Construction and Analysis

We have a ready-made program available in the form of the flowsheet and the specifications. The problem now is to change the written word into an electrical function. And, to make matters more complicated (or simple, depending on point of view), we only have ON or OFF to choose from. To quote an old saying, "The longest journey begins with the first step." Remember this! When a highly complex control system is required, or requires analysis, do not try to evaluate the whole system at once. Back-step to the very first functional operation. This usually consists of "press start button." From this beginning, the next steps can be taken, but one at a time—this keeps it simple.

10.15 The Rewrite

Before considering relays and switching devices we must rewrite the specification and flowsheet requirements in the language of logic. This approach may seem unnecessary and laborious, but it is the best approach. As we know from our chapter on mathematical logic (Chap. 19), the connectives AND and OR have very specific meanings. Consider first the AND. This is the logic product, and for a proposition to be "true," all the factors of the proposition must be "true." We use this AND in our rewrite to indicate an operational condition. If we have a number of motors $P1$, $P2$, $P3$, $P4$, and the specifications require that $P1$ AND $P2$ OR $P3$ AND $P4$ run together, then in the rewrite, we would state, Run $P1$ AND $P2$ OR $P3$ AND $P4$. This now specifically states that we have a choice of running $P1$ and $P2$; or alternatively, we can run $P3$ and $P4$; or alternatively, we can run $P1$ and $P2$, $P3$ and $P4$, all at the same time. The OR is normally the inclusive OR, meaning A or B or A and $B = T$. We immediately see that we encounter a problem of interpretation. Does the development engineer require that $P1$ and $P2$ run with $P3$ and $P4$ shut down? Or is it his intent to allow all four to run if required? For the four motors to run independently $(P1 + P2 + P3 + P4)$, the possible combinations of run and not run are 2^n, or, in this case, 2^4. Therefore, $2 \times 2 \times 2 \times 2 = 16$ possibilities exist. And, this is only our

A	B	A+B	
T	T	T	Both groups running
T	F	T	One group running
F	T	T	One group running
F	F	F	Both groups stopped

Fig. 10.4 Example 1—truth table for $A + B = ?$ Four combinations for running four motors in two groups of two. T is motors running, F is motors stopped.

first step. For the original arrangement of the four motors, the possible combinations are $P1$ AND $P2 = A$, and $P3$ AND $P4 = B$. As the motors are in two groups, we consider only possibilities concerning two.[1]

We do this by forming a simple table,[2] e.g., Fig. 10.4. One vertical column is used for each parameter. The number of horizontal columns

---

[1] We assume that the AND function is prewired for this example; otherwise, we use a truth table, as in Example 2, but with the problem $P1 \cdot P2 + P3 \cdot P4$.

[2] Sometimes called a *truth table*.

is 2^n, where n is the number of parameters. In our case, we have $n = 2$. This means two vertical columns and $2^2 = 4$ horizontal columns.

Notice that in the column under A parameter we have T, T, F, F. This is significant when making up a table. In the first column (from top to bottom), $2^n/2$ is T and the other half, F. In the next column, $B - 2^n/4$ is T, then $2^n/4$ is F, then $2^n/4$ is T and $2^n/4$ is F. As more parameters are used, more horizontal lines are required. The best way to explain the combinations will be to give another example. If we were considering the four pumps as $P1$ OR $P2$ OR $P3$ OR $P4$, we would have four parameters giving 4 vertical and 2^4 horizontal columns for all possibilities of ON/OFF arrangement. We set up a simple truth table in Fig. 10.5 with $2^n = 2^4 = 16$ horizontal columns and four vertical columns.

Notice that in vertical column $P1$, we have 16 horizontal lines. Working from top to bottom, we show $16/2 = 8$ at T, alternating with $16/2 = 8$ at F. Now, in the next vertical column $P2$, we work from top to bottom, getting $8/2 = 4$ and alternating with 4 at T and 4 at F. In the third column, we have $4/2 = 2$ and alternate with 2 at T and 2 at F. In the fourth column, we have $2/2 = 1$ and alternate with 1 at T and 1 at F. This completes the possibilities.

Now, we solve the problem of $P1 + P2 + P3 + P4 = ?$ We consider horizontal line 1 and see that we have $T + T + T + T = ?$ By laws of logic sums, we see that the valence is T. Knowing that in a sum only one addend need be true for a true proposition, we can now work down the columns inserting T, that is, until we get to line 16, where we have $F + F + F + F$; this equals F, therefore, 16 conditions. By extending the vertical columns we can also analyze the combinations for the original condition. This is without assuming that the pairs are, in fact, always T.

	P1	P2	P3	P4	P1 + P2 + P3 + P4 = ?	P1·P2 + P3·P4			= ?*
1	T	T	T	T	T	T	T	T	T
2	T	T	T	F	T	T	T	F	T
3	T	T	F	T	T	T	T	F	
4	T	T	F	F	T	T	T	F	
5	T	F	T	T	T	T	F	T	T
6	T	F	T	F	T	T	F	F	
7	T	F	F	T	T	T	F	F	F
8	T	F	F	F	T	T	F	F	
9	F	T	T	T	T	F	T	T	
10	F	T	T	F	T	F	T	F	
11	F	T	F	T	T	F	F	F	
12	F	T	F	F	T	F	F	F	
13	F	F	T	T	T	F	F	T	
14	F	F	T	F	T	F	F	F	
15	F	F	F	T	T	F	F	F	
16	F	F	F	F	F	F	F	F	

Fig. 10.5 Example 2—truth table for (a):
$$P1 + P2 + P3 + P4 = ?$$
Truth table to solve for (b):
$$P1 + P2 + P3 + P4 = 16 \text{ possibilities.}$$
$$P1 \cdot P2 + P3 \cdot P4 = 4 \text{ possibilities.}$$
T is motor running, F is motor stopped.

The table in Example 2 is spread out for simplicity's sake, but ordinarily the columns would be closed up and normal squared paper could then be used.

If we consider that each piece of equipment and every functional operation requires one relay, a small control circuit of 10 functions or relays had a 2^{10} arrangement and "open or closed" possibilities ($2 \times 2 \times 2 \times 2 \times 2 \times 2 \times 2 \times 2 \times 2 \times 2 = 1{,}024$). This gives 1,024 possibilities. This, in itself, is not a big problem with our single-step approach. However, after the circuit has been designed, a change in "one" step can sometimes cause matters to snowball into a complicated problem. What happens is that we must insert the change in the program and then review the whole rewrite of the program following the change. The change usually influences the whole sequence, and the review is more prone to error than a new step-by-step design. The rule is, then, before considering electric circuit layout, establish that the program is firm.

10.16 Logic Connectives

In the program rewrite, the grammar must be modified using the logic connectives AND, OR, THEN, and NOT. These four connectives should be the only ones used. If an exclusive OR is required, it should be written a and not b; b and not a. The word "then" is in actual fact understood as the connective for the implication. If we recall our mathematical logic, we remember that the implication takes the form, If a THEN $b = T$. We already assume that the proposition will be valid, and so when we consider the implication, we are considering a sequence; i.e., If this happens, then this must happen. This would be the first law of implication, that is, $T \rightarrow T = T$, and the law of *modus ponens*.

If the proposition is true, and the premise is true, then the conclusion must be true.

Now, in our program—wherever we have a sequence—we will use the connective THEN.

To recap briefly:

1. Operations simultaneously use the AND.
2. Operations A or B or both use the OR.
3. Operations A or B, but not both, use A AND NOT B, OR B AND NOT A.
4. Sequence of operation uses A, THEN B, THEN C, etc.
5. For lockout operation, use A AND NOT B.
6. Review of chapter on mathematical logic (Chap. 19).

10.17 The Circuit Layout

When the program is *firm*, and not before, we can begin the control-circuit layout. Before commencing, the symbols to be used should be shown on a sheet. For this chapter, we will use the symbols given in Fig. 10.6.

Normally closed push button	o⊥o	Normally open time closing—after 10 sec	TC 10 sec
Normally open push button	⊥ o o	Normally open limit switch	
Normally closed contact	⫫	Normally closed limit switch	o⟵⊸
Normally open contact	⊣⊢	Relay coil	(R)
Normally closed opening—after 10 sec	TO 10 sec	Indicating light— Red (R), Green (G), Amber (A)	(R)⊗

Fig. 10.6 Symbols.

The next step is to draw two vertical lines approximately 10 in. (optional) apart, and to show the first relay, as in Fig. 10.7. We now

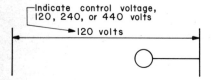

Fig. 10.7 Beginning a control schematic.

consider the first item on the program, and we assume a shutdown plant on start-up. This means that the circuit would be in the state when the plant power is OFF. The design is then approached as if you were the operator. Mentally, the conversation would go like this: "I press the start button, and motor *P*1 starts." Now, we show in Fig. 10.8 the start for *P*1 and also a STOP.

10.18 STOP/START Circuit

In this case, *P*1 is a magnetic motor starter, but the operating coil and starter function are similar to those of a relay. If the N/O START button is pressed, *P*1 will close. When the START button is released, then *P*1 becomes deenergized. To prevent this, we install a "seal-in" contact *P*1-1, across the START button. When *P*1 is energized, the *P*1-1 contact closes, so that when the START button is released,

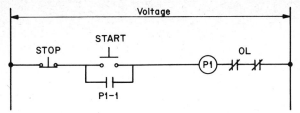

OL identifies overloads. Two or three are optional for three-phase motors.

Fig. 10.8 Standard STOP/START circuit and undervoltage release O.L. identifies overloads. Optional 2 or 3 for a three-phase motor.

$P1$-1 keeps the circuit energized. To stop the motor, the STOP button is pressed; this deenergizes $P1$, which causes $P1$-1 to open. When the STOP button is released, it once again closes the circuit. However, the $P1$-1 contact is now open: therefore, the circuit is interrupted at this point. This is known as *undervoltage release* and prevents the motor from automatically restarting after a power failure.

10.19 HAND/OFF/AUTO

Now, consider a second motor. If the program says $P1$, then $P2$, we require $P2$ to start as a result of $P1$ starting. For this, we use the HAND/OFF/AUTO switch. We show the diagram in Fig. 10.9 and add the next step. Notice that the contact number for the AUTO position is $P1$-3. This indicates a contact on the $P1$ starter. It also means that $P2$ cannot start until the $P1$ starter has been energized.

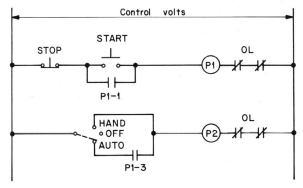

Fig. 10.9 STOP/START HAND/OFF/AUTO with $P1$ THEN $P2$ sequence.

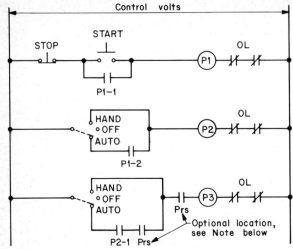

Note: Would not be located in HAND circuit unless it is a mandatory interlock under all conditions.

Fig. 10.10 Schematic for $P1$ THEN $P2$ and if pressure at Prs THEN $P3$.

10.20 Control Switches

As mentioned earlier, there are so-called digital control switches, which operate as a result of flow, pressure, temperature, etc. This must also be included in the control schematic. Think of a sample program as an extension of the program in Sec. 10.19: $P1$ then $P2$ and if pressure is at Prs, then $P3$ will run. The continuation of the schematic is shown in Fig. 10.10.

Note that the pressure switch Prs is located in the AUTO leg of the control circuit. This means that the HAND position will operate the motor without pressure. Alternatively, if we use the optional location, the HAND control will not start the motor unless there is pressure. This option should be clarified and included in the program.

10.21 Logic Variables

In symbolic logic, we have the logic calculus, which (like its numerical counterpart) deals with varying functions. It is not our intention to get involved in logic calculus, but we will use two of the definitions from the calculus. These are the definitions of the words "all" and "some." When we are dealing with a variable quantity of conditions, we state that we are considering "all" situations OR (exclusive OR) "some" of the situations. By definition, when we use "all," we mean "without exception." If there is at least one exception, then the word "some" must be used. Consider the following expression:

"All that glitters is not gold."

To rephrase this, we can say

"Without exception, everything that glitters is not gold, even a shiny gold wedding ring."

We can now see that the original statement, i.e., "All that glitters is not gold," is obviously false when it is taken literally. To be logically correct, it should be

"Some things that glitter are not gold."

It somehow ruins the Shakespearean[1] sageness of the statement, but it makes it logically correct.

10.22 Control-switch Option

Now, considering the various control switches, and the optional location by using the logic variable ALL or SOME, we can be specific in the program. To rephrase the program in Sec. 10.20, we would change from

1. $P1$ then $P2$ and if pressure is at Prs, then $P3$ will run.
2. $P1$ then $P2$ and (under all conditions) if pressure is at Prs, then $P3$ will run.
3. $P1$ then $P2$ and (under all conditions, except HAND operation) if pressure is at Prs, then $P3$ will run.

10.23 The Auxiliary Relay

The typical motor starter has normally only one auxiliary contact, which is used for "seal-in" contact. If more than one function is required, more contacts are necessary. The procedure, in this situation, is to use an auxiliary relay with multiple contacts. One of the contacts on the relay will then be used as the seal-in contact.

Consider the schematic in Fig. 10.11. We see in the schematic that if we start $P1$, the auxiliary contact $P1$-1 energizes relay $R1$. This closes contact $R1$-1, which seals in $P1$. At the same time, it opens the $R1$-4 contact. This locks out $P2$, making it impossible to start with the START button; the operation, then, is $(P1 = P2)'$, which is the logic equation for running $P1$ or $P2$, but not both together.

---

[1] Shakespearean type of quotation, not necessarily originated by Shakespeare.

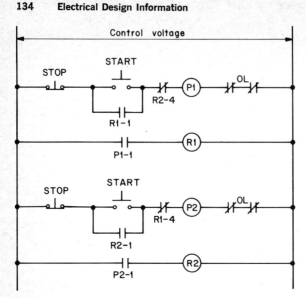

Fig. 10.11 Exclusive OR operation for $P1$, $P2$, using auxiliary relays.

The auxiliary relay can be sequenced to provide an infinite number of contacts. If we require 55 contacts to operate sequentially, we can use eight 8-pole relays: one contact for a seal-in and one sequence contact for $R1$ and one sequence contact for the remainder with one spare on $R8$, that is, $6 + 7 + 7 + 7 + 7 + 7 + 7 + 8 = 56$.

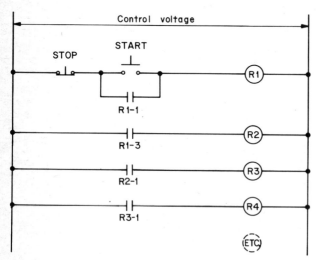

Fig. 10.12 Sequence operation for multiple relays.

10.24 Multiple STOP/START

Invariably, it is required that a motor have more than one location for stopping and starting. A STOP/START, or HAND/OFF/AUTO, is always located at the motor. Sometimes a STOP/START is included in the starter or motor control center. In designing a circuit with multiple STOP/STARTS, the rule is, *Connect all STOPS in series and all STARTS in parallel.* See Fig. 10.13.

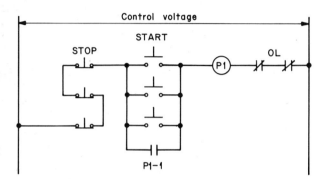

Fig. 10.13 Three STOP/STARTS with undervoltage release.

10.25 STOP/START and HAND/OFF/AUTO

Another situation arises where the control at the motor is a HAND/OFF/AUTO and the normal STOP/START(s) is (are) located remote from the motor. The connections, in this case, would be as shown in Fig. 10.14. Usually, the HAND/OFF/AUTO is intended for

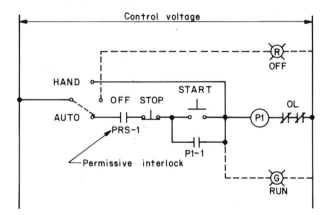

Fig. 10.14 Interlock controlled STOP/START circuit with HAND/OFF/AUTO at the motor and HAND override control. Optional indicating lights.

maintenance purposes or for removing a motor from the system. If we consider the preceding schematic, we would probably find that this application would be suitable for a long conveyor or a pump located at a tank, remote from the loading area. The maintenance man can test the motor by overriding the STOP/START and pressure switch.

The motor may be locked out of service by the HAND/OFF/AUTO being padlocked in the OFF position. If an indication is required that the motor is out of service, an indicating light may be connected to the HAND/OFF/AUTO OFF position contact. This will indicate that an inoperative system is due to removal from service, rather than an open pressure switch.

10.26 Undervoltage Release

If we consider the standard STOP/START circuit, we will see that if a power failure occurs, the motor-starter coil P_1 becomes deenergized. This causes the seal-in contact to open. When power is resupplied to the circuit, the motor starter remains deenergized. The motor must then be restarted with the START button. This is sometimes referred to as *undervoltage protection.*

Here, we must point out that "undervoltage protection" refers to the safety related to the unexpected start-up of a motor. It should not be confused with low-voltage protection, as supplied by a protective relay.

10.27 Summary

Designing control circuits is a matter of patience rather than of complexity. One of the main reasons for a long time being spent on the design of so-called complex circuits is the insufficient information supplied to the designer. Also, the designer usually "jumps in feet first," without an adequate program. This usually results in extra work and more probability of error.

There are numerous brochures issued by nearly all major electrical companies showing the design of standard control circuits. It must be emphasized, however, that the majority of control circuits are simply "build-ups" and modifications of the basic STOP/START HAND/OFF/AUTO principle. It should also be noted that in many cases the project engineer requesting the design of the circuit provides inadequate information in the first place. This may be due to lack of information on his part also.

Unfortunately when designing a control circuit there are no "maybe" or "not sure" operations. The "chips are down" and an exact operation must be programmed.

If it seems highly improbable that all the information will be available, make up the program and make a decision on the optional items. Next, issue the program for approval. A customer may not know what he wants, but he invariably knows what he does not want. Thus out of false information will come accurate information. Hopefully the program will be returned marked up with corrections or approved.

Chapter Eleven

Circuit Logic

11.1 Mathematical Logic

In Chap. 19, we discuss an unusual branch of mathematics. This is loosely termed *logic*. Very briefly, logic is an attempt to treat the interpretation of the written word as an exacting science of mathematical manipulation. Great accomplishments have been made, but many more refinements are necessary before it can be truly accepted as being developed to the level of normal algebra or the calculus.

One enormous problem is to make everyone familiar with the "one" and "one only" set of operational signs. For instance, the plus sign (+) (for this book) represents the inclusive OR. But for every book using this definition, there will be another book using plus (+) to represent the conjunction AND. The manipulations are basically established, but an electronics engineer will probably not recognize $(a = b)'$ as the equation for an exclusive OR. He probably will not be familiar with the "implication."

The converse is also probably true; i.e., the pure mathematician would probably not be familiar with the inversion or NAND gate, NAND being a NOT AND proposition, which is a negated product. However, a choice of systems is better than no systems at all. We will, therefore, stay with our own system and point out that when a different system is encountered, the ground rules must be studied because the probability is that signs and symbols will vary in meaning.

11.2 Circuit Logic

By applying the principles of mathematical logic to circuit analysis, we can sometimes establish a better procedure for circuit organization than the intuitive approach used in our "one-step" method. Our suggestion here is to use one system to complement the other. The brain is a far better computer than any that have yet been built. Unfortunately, like any other computer, if it starts with incorrect information, it can produce an incorrect answer.

With this in mind, use circuit logic as a tool to organize the system vagaries. After this, the intuitive one-step approach is usually adequate.

11.3 "Permissive" or "Hindrance" Solution?

When we are solving a logic network equation, we can take either of two choices. First, we consider the *permissive* solution, where a continuous flow is necessary to satisfy the network equation. For example:

$$a \cdot b = D$$

In the circuit in Fig. 11.1 the contacts must close for the relay D to operate. The contacts, then, must be permissive or GO to satisfy the AND proposition.

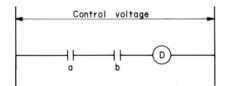

Fig. 11.1 Permissive solution $a \cdot b = D$.

Now, consider a similar circuit, but solve for a *hindrance* solution:

$$a' + b' = D$$

In the circuit in Fig. 11.2 we have a relay which is constantly energized. We then state that contact a' or b' must open to deenergize the circuit. Here we have two contacts in series, but we use an OR operator. The two examples show that we have two distinct methods of analysis, either one can be used. They are not generally interchanged. It is recommended that one system be used exclusively, preferably the permissive system. Once this becomes instinctive, then the hindrance system can be learned. In this book, we will only consider the permissive system, but application of De Morgan's law will indicate that the transfer from one system to the other is valid, and also simple.

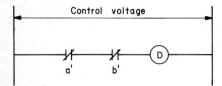

Fig. 11.2 Hindrance solution $a' + b' = D$.

11.4 The Logical AND

The AND is the logical product operator. All factors must be "true" for the proposition to be "true." We use the AND for the series-type circuits; see Fig. 11.3. For relay D to operate, contacts a, b, and c must be closed. The logic equation, then, is

$$\text{Relay } D = a \cdot b \cdot c$$

For relay D to operate, contacts a, b, and c must be closed; however, this is not their "off the shelf" state. By definition, then, we will indicate a normally open contact (N/O) with the *variable*. We will also show the normally closed contact (N/C) as the *negated variable*.

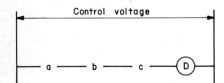

Fig. 11.3 AND circuit.

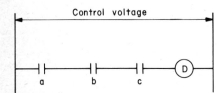

Fig. 11.4 Relay $D = a \cdot b \cdot c$.

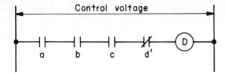

Fig. 11.5 Relay $D = a \cdot b \cdot c \cdot d'$.

The circuit for relay D is shown in Fig. 11.4. An alternative circuit is given in Fig. 11.5. The differences in the contacts then are

1. A change of state in a variable will cause the contact to complete the circuit.
2. A change of state in a negated variable will cause the contact to open the circuit.
3. The logical AND contacts are all arranged in series, regardless of whether it is the variable or a negated variable when solving for a permissive (GO) system.

11.5 The Logical OR

The logical OR is for the logical sum. If one or more variables are true, then the proposition is true. The conditions necessary for relay D to operate are shown in Fig. 11.6. Any contact closing will cause relay D to operate. Alternatively, all contacts closing will still

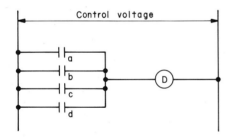

Fig. 11.6 Relay $D = a + b + c + d$.

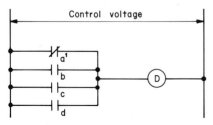

Fig. 11.7 Relay $D = a' + b + c + d$.

cause relay D to operate. The OR, then, requires all contacts in parallel. Any negated variable would cause relay D to remain closed all the time in its "off the shelf" position, for example, N/C contact a' in Fig. 11-7.

11.6 De Morgan's Law

To manipulate negated quantities within aggregators, we must use De Morgan's law. It is a very simple transfer manipulation:

$$(a \cdot b)' = a' + b'$$

Alternatively

$$(a + b)' = a' \cdot b'$$

In both cases, we negate *all* the variables within the aggregators and change the operational sign from AND to OR or vice versa.

Consider the circuit shown in Fig. 11.8. We have these windows as part of a burglar alarm system. The contacts are in series, and so to energize relay D, all contacts must close. This gives the circuit in Fig. 11.9. Now we see that in Fig. 11.9 if all the windows are set

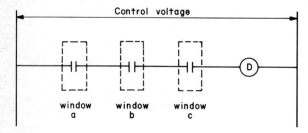

Fig. 11.8 Part of a simple burglar alarm system.

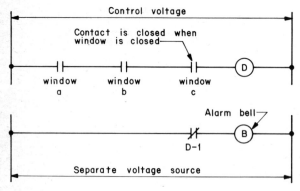

Fig. 11.9 $D = a \cdot b \cdot c$.

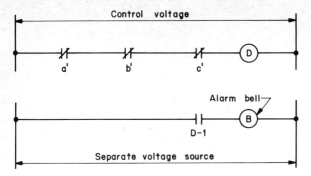

Fig. 11.10 $a' + b' + c' = B$. If a or b or c opens, then (B) alarm bell will ring.

and operate correctly the contacts then become closed and the circuit appears as in Fig. 11.10. The bell B is *not* energized because relay D is energized and the windows are in an energized (closed) condition. If we write the equation with reference to B and contact status, we have

1. $a \cdot b \cdot c = B'$ No alarm
2. $T \cdot T \cdot T = T$ Correct status

Now if we wish to explain what must happen for the alarm to sound we use De Morgan's law and change the status. This gives lines 3 and 4:

3. $(a \cdot b \cdot c)' = (B')'$
4. $a' + b' + c' = B$

In line 4 we have very subtly changed our viewpoint. This is sometimes difficult to understand. In line 1, we have a permissive circuit which in effect states, For B not to operate we must allow contacts a, b, and c to close. And in line 1 this is exactly what we did with $a \cdot b \cdot c = B'$. However, we are speaking in a negative framework. If we wish to speak in more normal language, we "invert" the whole viewpoint.

Therefore, what was once the operable condition requiring permissive action now becomes the "inert" condition and requires a "hindrance" or stoppage to initiate the action.

If we evaluate line 4 we will see that we have exactly this condition.

For B to operate we must cause a hindrance or stoppage of the energizing force and *not allow* a or b or c to stay closed; therefore, $a' + b' + c' = B$. In this last example, we showed the application of De Morgan's law. Remember that if the valence of a proposition is changed, that is, $T = T$ to $T' = T'$, then each side of the proposition should be enclosed and then negated. For example,

$$a + b + c = T$$

then

$$(a + b + c)' = T'$$

11.7 The Exclusive OR

The exclusive OR is shown by the logic equation $(a = b)'$. The proof is shown in the mathematics section. We can, therefore, state that

$$(a = b)' \text{ equivalent to } (a \cdot b') + (a' \cdot b)$$

In our logic circuits, we can only use the connectives AND and OR. Any equation in any other form must be modified to a product-sum form.

The exclusive OR function would be used where we have a choice of one or the other, but not both. Consider the circuit shown in Fig. 11.11. (In Fig. 11.11 Y_{LN} states that either A or B but not both can be energized at one time regardless of whether a_{PS} and b_{PS} are closed singly or nearly simultaneously.) In the circuit in Fig. 11.11, we see that if pressure switch a closes, it energizes relay A, which in turn opens contact $(A\text{-}1)'$. As the contact is normally closed, we show it as a negated contact. If we decided to include it in our equation, we would have

$$[a = A \cdot (A\text{-}1)' \cdot B'] + [b = A' \cdot B \cdot (B\text{-}1)']$$

Even though we use $A \cdot (A\text{-}1)'$, we are not stating that A and A-1 are the same item. Item A is the relay coil; item A-1 is a contact

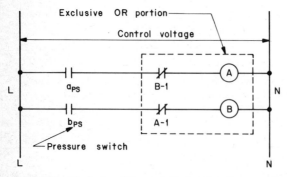

Fig. 11.11 Exclusive OR control circuit. Between L and N we have $a_{PS} + b_{PS}(A \cdot B' + B \cdot A') = Y_{LN}$.

which functions as a result of item A. If we show this in the form of an implication, we state

1. If coil A operates, then A-1 will operate also
2. If $A \rightarrow A$-1   Form implication
3. $A' + A$-1   Normalize
4. $(A' + A$-$1)' = A \cdot (A$-$1)'$ De Morgan's law
5. $A \rightarrow A$-1 equivalent to $A \cdot (A$-$1)'$ From lines 2 and 4

Thus in line 5, we show that $A \cdot (A$-$1)'$ is a valid method of describing the relay and its N/C contact.

11.8 The AND–OR Circuits

A circuit can have many combinations of contacts: series, parallel, series-parallel, and negated contacts. This does not pose any problem; we simply combine the various contacts. For example, the following arrangement gives the circuit depicted in Fig. 11.12.

$$a \cdot b \cdot c + d \cdot e \cdot f + g \cdot h' = D$$

We can also show that an unfamiliar arrangement can be misleading, because it does not appear in its normal form. Consider the circuit shown in Fig. 11.13. If we examine that circuit closely, we will see that it is a simple, parallel circuit. It is shown in standard arrangement in Fig. 11.14. Further, consider the circuit in Fig. 11.15. If we redraw the circuit, making the "jumper" infinitely small, we have the circuit shown in Fig. 11.16. This is more obviously recognizable.

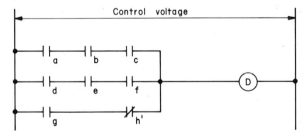

Fig. 11.12 $D = a \cdot b \cdot c + d \cdot e \cdot f + g \cdot h'$.

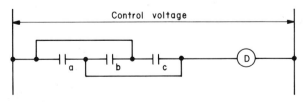

Fig. 11.13 $D = a\,?\,b\,?\,c\,?$ What is it? Series, parallel, or?

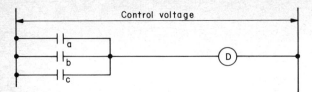

Fig. 11.14 $D = a + b + c$.

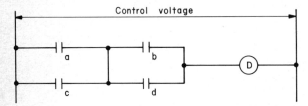

Fig. 11.15 What is it?

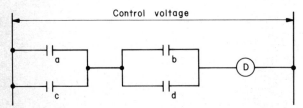

Fig. 11.16 $D = (a + c) \cdot (b + d)$.

11.9 Networks

A network is an arrangement of switches which, by their quantity and layout, require a systematic analysis. If we consider any network, it must eventually be traceable back to its power source, thereby making any network a "two-terminal" arrangement. If a network is multi-terminal and is analyzed at various points, then it will be subdivided into two-terminal sections. Then the subsections will be analyzed, and so on. A two-terminal network is shown as having a single input terminal m and a single output terminal n. We will use the letter Y for solving for permittances and the letter Z for solving for hindrances. For the following network, we will derive the solution

$$Ymn = a \cdot b \cdot c + d \cdot (f + g)'$$

This particular equation states that the network between terminals m and n, when solved for permittance, will give the logic equation as shown. It should be simple to construct the network from the above equation.

First, we rewrite the last part $(f + g)'$ to remove the aggregators. By De Morgan's law, we have

$$(f + g)' = f' \cdot g'$$

Therefore

$$Ymn = a \cdot b \cdot c + d \cdot f' \cdot g'$$

The results are shown in Fig. 11.17. Ymn is presented in actual form in Fig. 11.18.

Now consider a bridge-type circuit, as in Fig. 11.19. If we consider each path individually, as in our previous one-step approach, it should be relatively simple. Considering each path as a series circuit, the same variable or "gate" may be reused. We will then have optional paths, which become parallel circuits:

1. $m \cdot a \cdot b \cdot n$
2. $m \cdot a \cdot e \cdot d \cdot n$
3. $m \cdot c \cdot d \cdot n$
4. $m \cdot c \cdot e \cdot b \cdot n$

This, then, is the maximum number of possibilities. If we form the products into sums, we have

$$Ymn = a \cdot b + a \cdot e \cdot d + c \cdot d + c \cdot e \cdot b$$

The m and n are deleted because they are not contacts. They are terminal points and as such may be joined together. The equation is shown in circuit form in Fig. 11.20a. And if we make all the m and n into a single terminal, we have the circuit depicted in Fig. 11.20b. If we show this in contact form, we have the circuit of Fig. 11.20c. Here, then, we have the equivalent bridge circuit set into a logic equation. However, we know from the bridge network we started

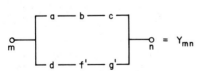

Fig. 11.17 Terminal m to $n = Y_{mn}$.

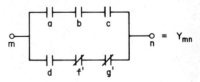

Fig. 11.18 $Y_{mn} = a \cdot b \cdot c + d \cdot f' \cdot g'$.

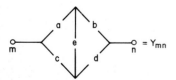

Fig. 11.19 Bridge circuit.

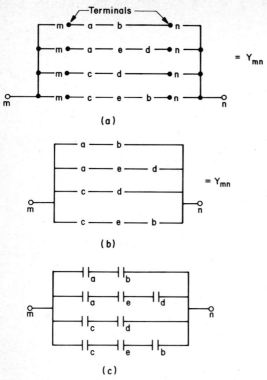

Fig. 11.20 (*a*, *b*) Bridge circuits; (*c*) equivalent bridge circuit.

with that all these contacts are not necessary. This leads us into the next section, on redundancy.

11.10 Redundancy and Absorption

As shown in the preceding paragraph, it is possible to expand or reduce the circuit components by manipulation and/or inspection. If we try to do it by inspection, we cannot be sure that we have considered all possibilities. Alternatively, if we set up our logic equation first, we can then factor out redundant components. Consider our last equation in Sec. 11.9:

1. $Ymn = m \cdot a \cdot b \cdot n + m \cdot a \cdot e \cdot d \cdot n + m \cdot c \cdot d \cdot n + m \cdot c \cdot e \cdot b \cdot n$
2. $Ymn = a \cdot b + a \cdot e \cdot d + c \cdot d + c \cdot e \cdot b$
   Here, we have removed the terminals m and n as
3. $Ymn = a(b + e \cdot d) + c(d + e \cdot b)$
   Factor line 2 (distributive law)

An Alternative Approach:

2a. $Ymn = a \cdot b + c \cdot d + a \cdot e \cdot d + c \cdot e \cdot b$
Rearrange line 2 (commutative law)
3a. $Ymn = a \cdot b + c \cdot d + e(a \cdot d + c \cdot b)$
Factor line 2a (distributive law)

We see that we cannot reduce the logic equation to its lowest form, i.e., with only a single a, b, c, d, e. This is because we have a delta formation, or, in fact, a double delta. If we recall the section on short

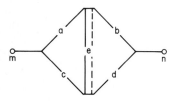

Fig. 11.21 Bridge circuit.

circuits (Chap. 9), we found that the delta arrangement did not lend itself to circuit analysis. We therefore converted to an equivalent star. The same applies to the logic equations as the state of the art exists today.

In lines 3 and 3a above, we see that we can factor redundant contacts except for the e. Here, then, is where the electrical or electronic Boolean algebra reaches its limitations.

With logic, however, we can prove the statement, We cannot represent the bridge circuit in logic form. At the same time, we can show that the same circuit can be analyzed so that we can state, Yes, it is a valid circuit.

First we will consider the actual circuit and try to represent it in logic form. We set up the circuit, using the program approach, and we have

$Ymn = a$ AND b OR c AND d OR if $a \cdot d$ THEN e, OR if b AND c,
$\qquad$ THEN e

1. $Ymn = a \cdot b + c \cdot d + (a \cdot d) \rightarrow e + (b \cdot c) \rightarrow e$
2. $Ymn = a \cdot b + c \cdot d + (a \cdot d)' + e + (b \cdot c)' + e$
3. $Ymn = a \cdot b + c \cdot d + a' + d' + b' + c' + e$
4. $Ymn = a \cdot b + c \cdot d + e$

Now, if we draw the circuit in line 1 and the circuit in line 4, we see that the logic equation in line 4 is false. We may now form another

150   Electrical Design Information

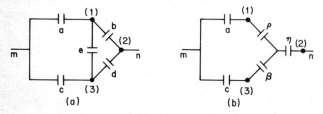

$Y_{mn} = a \cdot b + c \cdot d + (a \cdot d) \rightarrow e + (b \cdot c) \rightarrow e$

$Y_{mn} = a \cdot b + c \cdot d + e$

Fig. 11.22 (a) Bridge circuit; (b) not equivalent bridge circuit.

implication stating, If line 1, THEN line 4, OR line $1 \rightarrow$ line $4 = T$ or F. By substituting T or F for the premise and conclusion, we have $T \rightarrow F = F$. By our laws of implication (see Chap. 19), a proposition is always false if the premise is true and the conclusion is false. We evidently then can show the bridge circuit by using an implication, but we cannot reduce or standardize it any further with the logic definitions as they stand, although we have pointed out that we reduced the original equation to five contacts, which we could not do with our original approach in lines 3 and 3a (Sec. 11.10). This does not, however, make it a valid derivation within the rules of logic as they stand. However, we feel that the unidirectional characteristics of the implication could be useful if it were further developed for specific network analysis.

We now must still consider the validity of the bridge circuit and state whether it is a valid network or not. Even though we cannot represent the network in actual form, using ANDs and ORs, we can

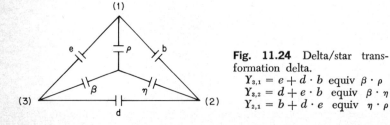

Fig. 11.23 (a) Original bridge; (b) equivalent bridge.

Fig. 11.24 Delta/star transformation delta.

$Y_{3,1} = e + d \cdot b$ equiv $\beta \cdot \rho$
$Y_{3,2} = d + e \cdot b$ equiv $\beta \cdot \eta$
$Y_{2,1} = b + d \cdot e$ equiv $\eta \cdot \rho$

set up an equivalent network and say yes or no to the circuit validity. In the bridge circuit, we have two parallel contacts and a single delta. We exchange the delta for the equivalent star. The rest is easy; see Fig. 11.23. Now, we determine the delta/star transformation and substitute; see Fig. 11.24. By substituting the star for the delta, we now have

$$Ymn = a \cdot \rho \cdot \eta + c \cdot \beta \cdot \eta \text{ (expanded form)}$$
$$Ymn = \eta(a \cdot \rho + c \cdot \beta) \text{ (distributive law)}$$

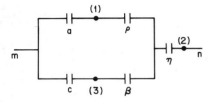

Fig. 11.25 Equivalent bridge circuit.
$Y_{mn} = a \cdot \rho \cdot \eta + c \cdot \beta \cdot \eta$ (expanded form).
$Y_{mn} = \eta(a \cdot \rho + c \cdot \beta)$ (distributive law).

Chapter Twelve

Wiring and Connection Diagrams

12.1 Control Diagram

When we are dealing with control circuits, we are interested in three different representations of the circuit(s). First, we have the *schematic*,[1] which shows the operating sequence. Second, we are interested in actual conduit and wire routing. Third, we need to know where to actually connect all the wires. The schematic diagram was covered in an earlier chapter but we did not show the method of numbering or identifying the wiring. We will also point out that there are many ways to actually arrange the drawing. We will show one method which is generally acceptable under all normal conditions. We will also show another method which is used but which is not always acceptable.

12.2 Schematic Diagram and Numbering

Consider the schematic diagram, the simple STOP/START circuit and light shown in Fig. 12.1. We see that the voltage is 120 volts for the circuit; therefore, we will use L_1 and N for control-circuit supply.

If we were using 240, 208, or 460 volts, then we would use L_2 instead

[1] Sometimes called *elementary diagram*.

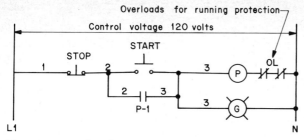

Fig. 12.1 Numbered wiring on a schematic.

of N for neutral. We do not show the control-circuit source and protection because there are numerous optional arrangements.

We then begin with number 1 at the left and we number left to right and down. Each time we encounter an operational device or contact the number will change except for pure terminal strips.

All wires which are spliced or connected to the same terminal should carry the same number. Whether the number is repeated on each wire is optional, depending on the circuit complexity.

All contacts, relays, lights, etc., must be numbered or identified. It is wise to establish permanent numbers or identifying letters at the start of schematic design. The actual wire numbering should be left until as late as possible to prevent unnecessary changes.

12.3 Wiring Diagram

Next we require information on the wireway and the number of conductors. We need to know the location of *all* control devices. These devices are all shown graphically in plan, in their relative locations. For example, consider the schematic in Fig. 12.2 with the pump and STOP/START location remote from the starter and with a green indicating light in a third location. By examination of the schematic, we see that wire 1 goes to the stop from $L1$, and so we show it. Wire 2 is from START/STOP to P-1, which is at P. We also have wire 3 from P to START. Last, we have a splice (3) at the T and N from source to G.

Now we are able to see how many conductors are required and how they are arranged in a wireway. The wireway and wire size are usually indicated on the conduit layout drawing. In fact, the relative locations are usually derived from the conduit plan.

12.4 Connection Diagram

We now have one more arrangement to make. This one is the connection drawing and shows the actual component connections. The dia-

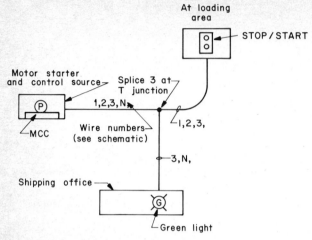

Fig. 12.2 Control (only) wiring diagram.

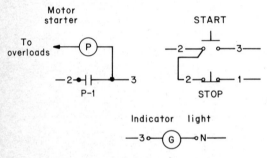

Fig. 12.3 Connection diagram.

gram for the same schematic appears in Fig. 12.3. We must now consider when this drawing will actually be used. First, the equipment will be installed, and the wires will be pulled in. The electrician has the numbered wire ends. At each component, he simply connects them as shown in the above diagram.

12.5 Wiring and Connection Diagrams Combined

A much simpler method of determining wire quantities is to combine the wiring and connection diagrams. The result is simply a diagram showing the component terminals, and then their interconnections. An example is given in Fig. 12.4.

If we show the components first and the wireway layout second and we connect each wire from point to point, we eventually finish with

Chapter Thirteen

Grounding

13.1 Definition

Circuits are grounded for the purpose of *limiting the voltage* upon the circuit which might otherwise occur through exposure to *lightning* OR *other voltages* higher than that for which the circuit is designed; OR, to *limit the maximum potential* to ground due to *normal voltage*.

If we analyze the above definition logically, we see that circuits are grounded to limit the voltage due to lightning a or other voltages b or to limit maximum potential due to normal voltage c. If we say that "limit the voltage" = "limit maximum potential," then we can factor this out and say:

"Circuits are grounded to limit the voltage due to $a + b + c$, where a, b, and c are lightning, higher voltages, and normal voltage, respectively."

If the definition is correct and our logic is correct, we can, by contraction, state:

"Circuits are grounded to limit the voltage."

This then will be our source of reference. If a problem exists where the guidelines are not clear, then the problem related to this definition should resolve itself.

13.2 System and Equipment Grounding

Basically, there are two types of grounding:

1. The grounding of systems, where an actual current-carrying part of the circuit is grounded
2. The grounding of equipment, where a non-current-carrying portion of the equipment is grounded

Although by our definition we are concerned with limiting the maximum voltage, the procedures and reasoning differ in the two types of grounding. We will consider system grounding first, and then equipment grounding.

13.3 System Grounding

We generally assume that systems are either *delta* or *star*. All other systems are either dc or single-phase offshoots of the basic system. We will consider only the three-phase delta, three-phase star, and one-phase systems. System grounding involves extensive study and requires that the engineer have a certain expertise. He must deal with such things as harmonic components, zero sequence components, and choice of resistor, reactor, or solid grounding. Where are the best grounding points? Do we ground the source or the load? And so on. We will point out the salient features of the systems, with the intention of introducing them briefly to the reader. When it comes to selecting specialized grounding equipment, then the manufacturer should be contacted, unless the reader happens to be an expert.

13.4 Delta System

The delta system, by virtue of its arrangement, does not have a neutral point common to all three phases. Therefore, it does not lend itself to grounding as readily as does the star system. An ungrounded delta system under fault conditions can cause high transient overvoltages. A ground fault on one phase can cause line-to-line voltages from each of the other phases to ground, which is $1.73/1 = 1.73$ PU (per unit) or 73% higher than normal. In some cases, the line-to-ground insulation can stand the overvoltage. If it does not, then a line-to-line fault will develop, whereas the original fault was a line-to-ground.

13.5 Delta System Ground

If it is decided to ground a delta system, then the alternative methods are

1. Corners-of-delta grounding
2. Star/delta transformer
3. Zigzag transformer

The *corners-of-delta grounding* will only provide partial protection and will probably create more problems than it will solve. We do not recommend this method.

The *star/delta grounding* is a solution. With a two-winding star/delta transformer, the star side of the transformer is connected to the line. The winding, therefore, must be for the same voltage as the system. The neutral of the star winding is connected to ground, either directly (solid) or through a resistance. The delta winding is closed and will allow zero sequence currents to circulate. The delta winding may also supply a normal load if required.

The *zigzag grounding transformer* is a single-winding transformer, but each phase winding is split into two halves. The first part of one phase is connected in series with the second part of the next phase and so on. One advantage of the zigzag transformer is that (due to

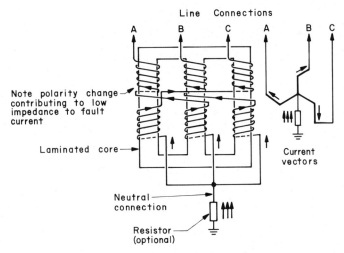

Fig. 13.1 Three-phase zigzag grounding transformer.

the phase interconnections) a ground fault on one phase will be equally distributed between all phases. Another advantage is that the construction parts are only 58% of the equivalent star/delta grounding transformer of equal kva. The possible use of the zigzag grounding transformer should receive first consideration. After the zigzag has been investigated, the star/delta can be compared for price and convenience, with a low-voltage, three-phase source as a possible bonus.

13.6 The Star System

The three-phase star (or wye) system is the obvious system of choice when one is considering systems which will be grounded. The neutral point may be readily grounded, which provides a three-phase, four-wire system with all phase conductors tied to ground potential. This reduces the transient overvoltage magnitudes. It also limits the overvoltage appearing on the other phases when a ground fault occurs on a single phase. Lightning protection is improved and ground fault locations are more obvious. Generally, there are many advantages to the grounded system over the ungrounded.

13.7 Solid, Resistance, Reactance Grounding

Each of the above three methods of grounding has a specific application depending on system parameters, source equipment, and reason for grounding. The solid ground is usually used in low-voltage systems, that is, 600 volts and below, and in high-voltage systems, over 14,600 volts. The resistance ground is usually applied to distribution voltages, that is, 2,300 to 14,600 volts. The reactance ground is used as an option to the resistance ground provided that the reactance value is low. The high-reactance ground is not recommended, due to possible overvoltage contribution.

13.8 Ground Location

From the utility company down to the low-voltage user, the voltage is gradually stepped down through possibly four or five steps. In a grounded system, this could be

    Delta/star   Delta/star   Delta/star
    500–220 kv   220–110 kv   110 kv–33,000 volts

    Delta/star   Delta/star
    33,000–12,000 volts 12,000–480 volts

At each voltage level, the star point must be grounded. Where a choice exists, it is preferable to ground a system at the source and not at the primaries of star/delta transformers. Each case, however, should be evaluated individually, rather than making a hard-and-fast ruling against it.

13.9 Low-voltage Systems

All low-voltage power systems such as 480/277, 208/120, and 240/120 with a neutral are essentially sources of single-phase power. As such, all neutrals should be grounded at the source.[1] This is ahead of the service-disconnecting means. No further connections shall be made to a grounding electrode on the load side of the service disconnect (see Fig. 13.2).

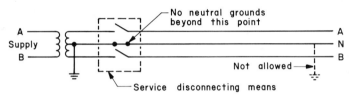

Fig. 13.2 Single-phase system showing allowable neutral grounds.

13.10 Identified Conductor

NEC 250-25 gives the details on "the conductor to be grounded." In all cases, the identified conductor should be white, including neutral conductors. The covering of large size conductors may be black or other than white. In this case, in any box or device in which the identified conductor appears, the identified conductor must be painted white.

13.11 Grounding Conductor

This conductor has the sole function of providing a continuous path to ground from non-current-carrying parts of electric equipment. This can be motor frames, switching equipment, transformers, etc. The grounding conductor may be bare copper or noncorrosive conducting material. If it is insulated, the color must be green or green with yellow tracer. Do not confuse the "grounded identified conductor" (white) with the "grounding conductor" (green or bare).

13.12 Equipment Grounding

In NEC 250-42 and 250-43, conditions are listed wherein fixed electric equipment, which may become energized, must be grounded. In an

---

[1] The power transformer.

industrial plant, it is hard to find a piece of equipment which should not be grounded. The general rule then is to ground all fixed equipment.

13.13 Enclosure Grounding

All enclosures, such as panels, switchgear, transformers, motors, etc., should be grounded. NEC 250-33 lists an exception; however, the conditions are such that rather than observe the exception, install a ground. Usually the grounding is accomplished by a conduit wiring system anyway.

13.14 Nonelectric Equipment Grounding

NEC 250-44 lists requirements for grounding of metal parts of nonelectric equipment. Essentially it points out that although in actual fact some equipment may be nonelectric, the possibility exists that its proximity to electric equipment would warrant its grounding purely in the interest of safety.

13.15 Portable Equipment Grounding

In an industrial plant, it is wise to install 110-volt, one-phase convenience receptacles of the grounded type. The three conductors into the receptacle are then phase, neutral, and ground, or black, white, and green. All portable hand tools should have the matching plug and cord to extend the ground to the tool.

13.16 Structural Grounding

Any electric equipment which is actually secured to and in contact with the *structural* metal frame of the building can be considered as grounded. We must point out that this applies only to structural metal used as *building support and framing*. It does *not* apply to a metal structure built for some other purpose adjacent to or on a floor of the building.

13.17 Lightning Conductors

Where lightning rod conductors come in close proximity to a ground conductor, i.e., within 6 ft, then they should be bonded together.

13.18 Bonding

Any device, equipment, or enclosure which forms a part of a continuous ground system and offers the possibility of poor continuity or high resistance should be bypassed and/or bonded with a jumper of corrosion-resistant material at every point where a possible discontinuity may occur. Typical examples could be wireways with insulated sections for radio frequency interference (RFI)[1] prevention, loosely jointed metal raceways, expansion joints, flexible motor connections, etc.

13.19 Grounding Source

A grounding conductor installation becomes useless if there is no true ground source to connect it to. Even bare copper wire buried in the ground is no guarantee of an acceptable ground. One of the first requirements then for a new installation is to examine the ground conditions. This should be done as soon as a site is located. Sometimes it is necessary to drill down very deep to obtain a suitable ground. This may involve a major cost item that the client had not considered. Therefore, do not assume that a good ground exists at a new site. Call out in the specifications for formal ground tests to be made.

13.20 Water Pipe Bonding

The first ground source to consider is a buried water pipe system. If there is a water meter in the line, then it should be bypassed by a bonding jumper. NEC 250-81 points out that the piping system could be disconnected in places or it could be disconnected in the future. Therefore, it is good practice anyway to add supplemental grounds at other points in the system.

13.21 Structural Steel Grounding

Although structural steel is acceptable as a grounding point, it is only as good as the ground conditions that the building rests on. Sometimes the design of the grounding system is completed before the building design. Therefore, the normal routine would be to establish a main ground source. Good practice would be to install a buried ground loop or main continuous ground conductor and bond the structural steel at various points. It is then permissible to use the structural steel to pick up various ground points within the building.

[1] When a condition like this occurs the raceway is grounded at one end only to eliminate closed loops.

13.22 "Made" Electrodes

"Made" electrodes is the term usually employed to refer to the supplemental grounds or a ground source established specifically in a remote location, which precludes the running of the main ground conductor over a long distance. NEC 250-83 describes the alternatives available.

13.23 Ground Wells

The term *ground well* usually refers to a standard detail showing the construction of a "made" electrode. This sometimes takes the form of driven rods with a protective soil pipe and cover. This prevents the possibility of the ground wire being accidentally broken or disconnected.

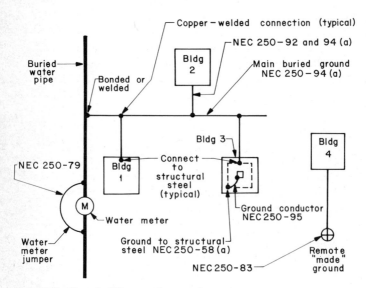

Fig. 13.3 Four buildings with ground system.

13.24 Ground Resistance

The resistance between the connection point of a "made" electrode, or in fact any main ground source, and ground should be no more than 25 ohms. In general, a water pipe system will usually be adequate. NEC 250-84 states that an underground water pipe will *in general* have a resistance to ground of under 3 ohms. The question is, How do we know that the conditions are "in general"? It is therefore necessary to test new site locations, and also to carry out periodic testing of existing plants.

13.25 Ground Testing

There are various methods of testing ground resistance. Due to the changing climatic conditions and varying density and conductivity of the earth, it is virtually impossible to make any general statements. A general statement is meaningless for specific applications, but it is useful as an indication of where to begin. It never precludes the specific information. We must accept the fact that a resistance test should be made.

When a test is made, test electrodes are driven into the ground at various points. If possible, this should be done 2 or 3 months ahead of the testing. This allows for the effects of aging. Regardless of the type of test circuit used, all the conditions of the test should be recorded. The test will provide a result that is only good for that set of conditions. These conditions must be very carefully outlined. The value of the test when it is applied to conditions other than the test conditions must be weighed against the basic test parameters. Then, a factor of probability enters the picture.

There are other circuits such as a normal Wheatstone bridge circuit. Also, there are various commercial test units specifically designed for this purpose. As long as the specific test parameters and the circuit used are outlined, the test data will be evaluated taking these factors into consideration. Do not get into the embarrassing position of running a test and having the client spend a few thousand dollars on drilling ground wells only to find that it was not necessary.

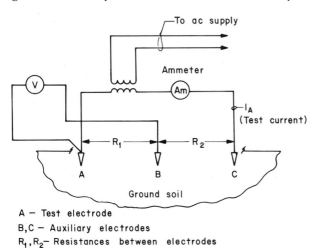

A – Test electrode
B,C – Auxiliary electrodes
R_1, R_2 – Resistances between electrodes
V – High-impedance voltmeter
Am – Ammeter

Fig. 13.4 Simple ground test circuit.

Chapter Fourteen

Lighting

14.1 The Art of Lighting

There are so many varied types of fixtures and light sources available today that designing a lighting layout allows a certain aesthetic latitude which was not formerly available.

Although the industrial-type plant is more concerned with adequate lighting than aesthetic qualities, it is a fact that layout, glare, heat, and color have a psychological effect on individuals.

Unfortunately, when a client submits his criteria, he will sometimes specify the footcandle levels for the various areas and the type of lighting to use. In effect, the art of lighting is generally ignored in industrial plants. Economy factors are weighed against all other factors. Usually, economy wins out over the optimum design.

In industrial plants, there are usually offices which have commercial-type lighting. This requires a different design approach than industrial lighting.

14.2 Lighting Design

Even though we know that lighting design can be an "art," we will consider the normal design practice. In offices or assembly rooms, where the general floor space is taken up with desks or assembly benches, then fluorescent lighting is the first consideration.

In areas where the space is "high bay" (high roof), with machinery taking up floor space, the trend is to mercury vapor lighting.

The incandescent-type lighting is usually relegated to closets, doorways, night lights, toilets, and local lighting on a machine. For example, a flexible gooseneck-shielded incandescent may be used on a lathe and directed onto the cutting tool to illuminate the cut.

14.3 Lighting Calculations

Clearly, then, we have two general classes of lighting (excluding floodlighting). One is the closed-room, low-ceiling type of installation. The other is the high-bay, high-roof, large-roof-span buildings, where walls may as well not exist as far as lighting is concerned. In fact, sometimes part of the manufacturing is outdoors, where there are no walls or roofs. Two different methods of calculation are necessary.

14.4 Lumen Method

This method of lighting calculation is for rooms. The basic equation is relatively simple:

$$fc = \frac{\text{lamp lumens} \times MF \times CU}{\text{area}}$$

where MF is maintenance factor, CU is coefficient of utilization, and fc is footcandles.

By cross multiplying, we can solve for any of the other units, e.g.,

$$\text{Lamp lumens} = \frac{fc \times \text{area}}{CU \times MF}$$

It is clear that the lamp lumens are fixed, depending on the fixture and lamps selected. The footcandles are fixed by specification. The area is also fixed by measurement. The only variables are the coefficient of utilization and the maintenance factor. The maintenance factor can be considered as 75% or 0.75 for good conditions with a low of 65% or 0.65 for poor conditions. Since one guess is as good as another, in trying to predict maintenance of a nonexistent plant, a figure of 70% or 0.70 should be a fairly low consistent constant.

Now we have only the coefficient of utilization to consider. This is the most confusing coefficient of all. It takes into consideration the shape of the room (room ratio). We also consider wall, floor, and

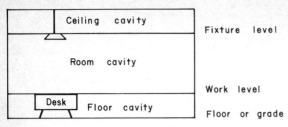

Fig. 14.1 Cavity measurements.

ceiling colors for reflectance values. The room ratio can be figured, but rarely is the color of walls or floor or ceiling known. Even if they are, a factor is assigned which varies between 0.75 and 0.10. Here again, the accuracy of the guess is open to question. A further recent development is the cavity ratio. This takes into account three measurements; see Fig. 14.1. These are variations of the room ratio measurements. By referring to tables, we get an end result from all these factors which is the coefficient of utilization. Regardless of what the tables predict, if we use a CU that is excessively high, there will probably be questions as to its validity. If we use one too low, it will require too many fixtures. A CU of 0.6 to 0.7 should be a fair assumption.

14.5 Validity of Lighting Calculations

In the preceding section, we have shown how to calculate for a certain lighting level. But what does this calculation mean? Essentially, we try to predict a time span. We state that if the lamps age, and the walls turn brown, and no maintenance is carried out, then the lighting will not drop below the calculated level. However, we do not state "in 6 months" or "in 2 years," etc. We are trying to use static quantities and apply them to dynamic functions. The chance of making a lighting calculation and measuring it at a predictable time, and of having the calculations match the light meter reading, is remote.

The calculations are not wasted, but they are only part of the picture. They give us a number, so that we can decide on a approximate number of fixtures. After that, we determine the arrangement. Now, do we use low light output and more closely spaced fixtures, or high output and more widely spaced fixtures? Consider also the glare and the type of fixture louver or diffuser, as well as the color content in the flourescent- and mercury-vapor-type lighting. The actual lighting layout with a pleasant-looking diffused fixture and a reasonable footcandle level is far more efficient than a high footcandle level with the glare of exposed tubes. Try to contribute some of the light to the ceiling (i.e., upward component). This eliminates a tunnel effect. In a hot area, try for cool-color lighting; in a cold area, try for the opposite effect. The amount of time usually allowed for the design of a lighting layout is not unlimited, and in fact any excessive amount of time spent may be a matter of concern, so that an optimum time should be allotted.

14.6 Rule of Thumb

A good approximation for power requirements for a lighting design is 5 watts/sq ft for incandescent lighting to obtain 50 fc, and 2 watts/sq ft for fluorescent and mercury vapor to obtain 50 fc. This is not intended to replace the lighting calculations; but if the lighting calculations do not compare with these rule-of-thumb figures, then recheck the lighting calculations.

Consider a room 40×20 ft, with fluorescent lighting required at 50 fc (see Fig. 14.2).

$$\text{Lamp lumens} = \frac{\text{fc} \times \text{area}}{\text{CU} \times \text{MF}} = \frac{50 \times 40 \times 20}{0.7 \times 0.6} = 95{,}238 \text{ lumens}$$

40-watt rapid start = 3,150 lumens
95,238/3,150 = 30.3 lamps; therefore use 30 or 32 *lamps*

We have a 4-ft-long fixture and two continuous rows on 10-ft spacing. Delete one fixture to allow 2 ft at each end. This gives nine fixtures per row. With 2 rows, there are 18 fixtures; and at 2 lamps per fixture, there are 36 lamps. This requires two extra fixtures over and above our original calculations. However, to eliminate dark spots at each end, they would be necessary. Now, by rule of thumb, check:

1. 800 sq ft at 2 watts/sq ft = 1,600 watts
2. Using 40-watt lamps, 1,600/40 = 40 lamps
3. Two lamps per fixture = 20 fixtures
4. Two rows gives 10 fixtures per row

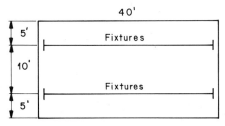

Fig. 14.2 40×20 ft room and fixtures.

5. 10 × 4 ft = 40 ft, which is too long, and so delete one fixture per row
6. This gives two-lamp fixtures per row
7. 18 + 18 = 36 lamps

We see, then, that we have exactly the same quantity of fixtures with our rule of thumb as we do with our calculations. It may not always work out exactly, but it is usually a reliable guide or check.

14.7 Fixture Spacing

In general, fixtures in office-type rooms and low-bay production assembly rooms are spaced around a 10-ft maximum area. The desired spacing may have to be adapted to available structural conditions. The approach, then, is to calculate the optimum spacing and then check for mounting conditions and modify accordingly. This will mean installing special mounting supports so as to maintain the desired spacing, or relocating to existing available supports.

For N rows of fixtures, we have $2N + 1$ dividing marks; see Figs. 14.3 and 14.4 for the location of fixtures beginning with one wall. Notice that for correct spacing the distance from the wall to a fixture is exactly half the fixture-to-fixture distance.

Fixture spacing and mounting height should be nearly equivalent. Consider the previous example of a room 40 × 20 ft at 50 fc. We decided on two rows of two-lamp fixtures each. Now, assume we had to redesign for 100 fc. We have a choice of two rows of four-lamp fixtures or $2 \times 9 \times 4 = 72$ lamps or three rows of three lamp fixtures

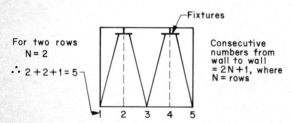

Fig. 14.3 Fixture spacing. Note that five numbers require four equal spaces.

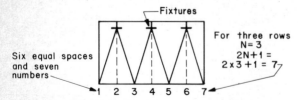

Fig. 14.4 Fixtures are located on even numbers.

giving $3 \times 9 \times 3 = 81$ lamps. With nine lamps difference, the decision is now probably one of economics.

Choice 1 is labor and cost of three rows of three-lamp fixtures.

Choice 2 is labor and cost of two rows of four-lamp fixtures.

In case 1, the labor cost will be high and the fixture cost slightly lower than that of a four-lamp fixture. In case 2, we have the inverse condition. The preferable case for good lighting would be the three rows of lower output. Here, then, someone must make the decision.

14.8 Louvers and Diffusers

As previously stated, there are numerous fixtures with more numerous design characteristics. Here, unfortunately, only familiarity with the lighting catalogs will help. There is usually a specific type of fixture that the designer visualizes. The procedure, then, is to select two or three of the desired type, and then to evaluate the costs.

14.9 High-bay Lighting

In industrial plants, we sometimes have areas with high roofs and huge floor space. These areas should not be analyzed by the lumen method. This is for the obvious reason that the coefficient of utilization is not valid and the fixture mounting heights are usually excessive. With high-bay lighting, more care is required when assigning footcandle levels. As mounting height increases, light reduces as the square of the distance.[1] If we move a fixture from 10- to 20-ft mounting height, we will get only one-fourth of the light. This is known as the inverse-square law. Very quickly, we can see that if we change an area from low-bay to high-bay lighting, possibly due to a crane installation, then we could conceivably require 4 times the number of fixtures to maintain the previous footcandle level. Thus, the cost per installed footcandle for high-bay lighting is more than the equivalent low-bay cost. Another higher item is the installation (labor) cost due to awkward scaffolding and to possible bonus rates for high-level working conditions. The rule, then, is to spend more time studying the footcandle requirements for a high-bay design than for a low-bay one.

14.10 Fixture Selection

High-bay lighting is usually mercury vapor or high-output fluorescent. The choice of fixtures is much more limited. High-bay lighting is a

---

[1] For light without lens.

compromise in design and economics. It is a rare customer who issues a blank check for high-bay lighting. When an installation involves large quantities of fixtures, it is good practice to call in a fixture manufacturer and discuss an economical installation. Also, the possibility exists that there may be new fixtures available but not shown in the catalog.

14.11 Candlepower Distribution Curve

If we measure the light output of a lamp, it will vary depending on the point it is measured from. If we assign 0° to a point directly below the lamp, then 180° must be directly above it. (See Fig. 14.5.) If we make a table giving candlepower readings every 10°, we have the information to plot a curve. This is good for a particular shape of fixture only. The fixture manufacturer will publish this *photometric data* in the catalog. This information is necessary for the calculations. The examples in Figs. 14.6a and b and 14.7a and b are from a manufacturer's catalog. We will refer back to these as we carry out our calculations.

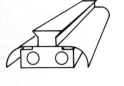

Industrial Fixture —
1,500-ma turret,
semidirect, with 2 lamps
VHO/215 w
(VHO— very high output)

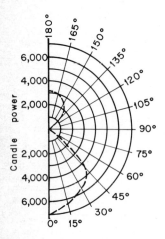

Degrees	C. Power
0	7,000
30	5,700
45	4,200
60	0
90	0
105	0
120	0
135	1,500
150	2,500
165	3,000
180	3,100

Measured values plotted against various angles

Fig. 14.5 Photometric curve and candlepower table.

Lighting 173

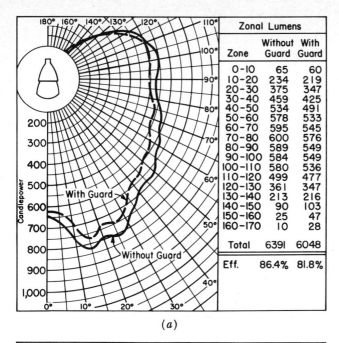

(a)

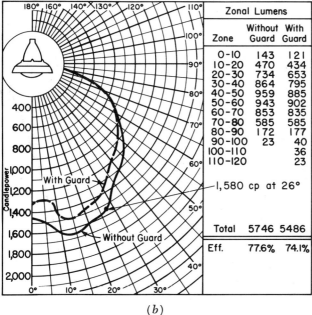

(b)

Fig. 14.6 DL series fixtures, MV (Hg) distribution curves.

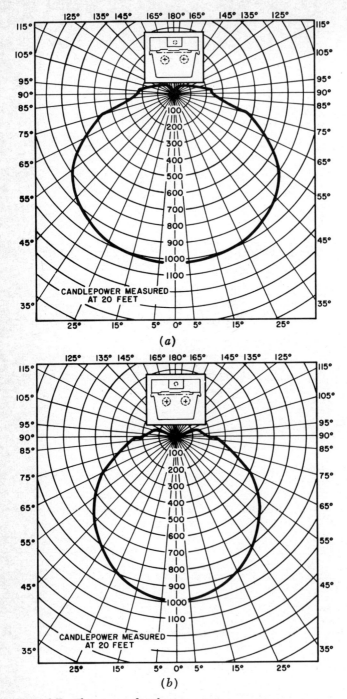

Fig. 14.7 Fluorescent distribution curves.

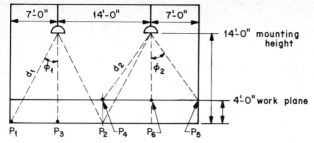

Fig. 14.8 Point-to-point diagram.

14.12 Point-to-point Calculation

With high-bay lighting, we consider an exact spot and figure the illumination at that point. We take into account the height of the workbench, the height of the fixture, and the fixture spacing. We begin by considering a minimum footcandle level. We next determine what the mounting height must be and the approximate spacing. The mounting height should remain firm, but the spacing between fixtures will be determined by means of the calculations (see Fig. 14.8).

$$fc = \frac{cp \times \cos\theta \times MF}{d^2}$$

where fc = footcandles
 cp = candlepower
 MF = maintenance factor
 d = hypotenuse

We have selected a fixture, and so now we evaluate the lighting level at various points. By selecting P_1 first, we must now determine $\cos\theta_1$ and d_1. By simple trigonometry and knowing the mounting height and spacing, we have (for DL series mercury vapor 175 watt) mounting height = 14 ft

1. $\tan^{-1} 7/14 = \tan^{-1} 0.5 = 26°$
2. $\cos^{-1} 0.9 = 26°$
3. $d_1^2 = (14)^2 + (7)^2 = 196 + 49 = 245$
4. From photometric data 26° gives a candlepower of 1,580 (Fig. 14.6b).
5. $fc_{P_1} = \dfrac{1{,}580 \times 0.9 \times 0.7}{245} = 4.0 \text{ fc}$

We have two fixtures supplying light to point P_2, so that

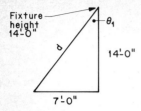

Fig. 14.9 A triangle formed by spacing and mounting height.

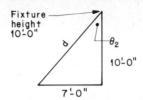

Fig. 14.10 Mounting height and spacing triangle.

where 1,425 is taken from the data curve in Fig. 14.6b, $\cos 0° = 1$, and $d^2 = (14)^2$ at $0°$. Point P_5 is at $14 - 4 = 10$ ft below the fixture and 7 ft from the center (see Fig. 14.10).

$$\tan^{-1} \theta_2 \; 7/10 = \tan^{-1} \theta_2 \; 0.7 = 35°$$
$$\cos^{-1} 0.8 = 35°$$
$$cp = 1,300 \; (\text{Fig. 14.6}b)$$
$$d^2 = 10^2 + 7^2 = 149$$
$$fc_{P_5} = \frac{1,300 \times 0.8 \times 0.7}{149} = 4.9 \; fc$$
$$fc_{P_4} = 4.9 \times 2 = 9.8 \; fc \quad (\text{due to two lights})$$
$$fc_{P_6} = \frac{1,425 \times 1 \times 0.7}{(10)^2} = 10$$

We now have a set of lighting levels to use as a base. We can now state that for the *same mounting height* for 20 fc, we require

$$cp = \frac{fc \times d^2}{\cos \theta \times MF}$$

For point P_2, see the P_1 calculation and cross multiply.

$$cp = \frac{20 \times 245}{0.9 \times 0.7} = 7,777 \; cp \; at \; 26°$$

However, this is for two fixtures, and so

$$7,777/2 = 3,888 \; cp \; at \; 26° \; per \; fixture \; (\text{Fig. 14.11}a)$$

We consider a similar fixture in Fig. 14.11a and b. With a 400-watt lamp, we obtain 4,150 cp output, which is slightly higher than what we require but is acceptable. We also note that as we approach 35°, the distribution curve does not change as rapidly on the new fixture as it did on the first one.

By juggling fixtures, mounting heights, and spacing, we eventually determine an economical installation.

Lighting 177

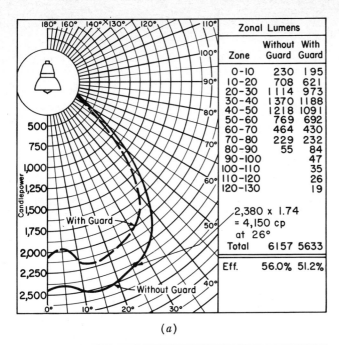

(a)

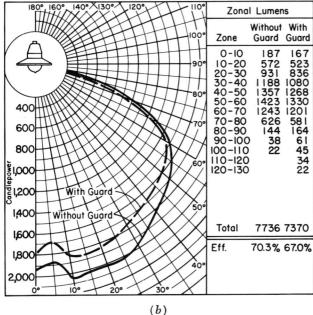

(b)

Fig. 14.11 Hg (mercury) distribution curves.

14.13 Circuiting

Industrial lighting is usually controlled by the circuit breaker instead of by a switch. If the circuit loading is heavy and a remote switch is necessary, it may require a lighting contactor. This device is an electromagnet operating a set of contacts, or basically an oversize relay. When long rows of fixtures are installed, every other fixture or row should be on alternate circuits. Single-phase circuiting is shown in Fig. 14.12a and three-phase, four-wire circuiting in Fig. 14.12b. Certain fixtures should be assigned to night lights and a special circuit assigned to control the night lights. Under certain conditions, single rows of fixtures can go on a single circuit. This would suit a condition where a product assembly area fluctuates. As extra space is required, another row of lights is turned on.

14.14 Floodlighting

Due to the changing times, the individual light over the doorway is being replaced by good outdoor floodlighting. The individual floodlighting fixtures have photometric data and usually the design information to predict lighting levels. We will not consider floodlighting design, as it will vary with the fixture. We will, however, consider the philosophy behind outdoor lighting. The eye is attracted by contrast. Therefore, when we design outdoor lighting, we are concerned with contrast, not lumen level. We know that there are more automobile accidents at dusk than in pitch darkness. This is due to a lack of contrast at dusk.

Depending on the area, then, we will light very brightly so that the

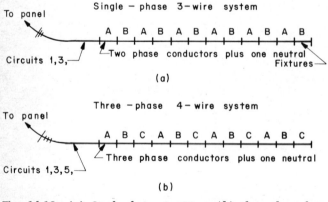

Fig. 14.12 (a) Single-phase circuiting; (b) three-phase four-wire circuiting.

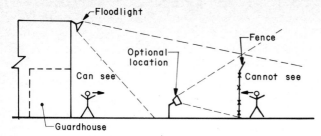

Fig. 14.13 Security lighting—low-level lighting.

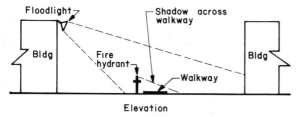

Fig. 14.14 Shadow across pathway.

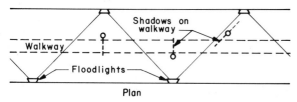

Fig. 14.15 Staggered floods for low-level lighting.

area is clearly illuminated, or we will light very dimly to create shadows from objects. The strategic location of outdoor floodlighting is thus of paramount importance. Security fence lighting should have the effect of creating a silhouette when an individual is viewed from the guardhouse. It should block an intruder's vision and yet allow the guard to clearly identify movement (see Fig. 14.13). When lighting passageways between buildings, locate the lighting so that shadows from obstructions fall across the walkway, as is shown in Fig. 14.14. In general, outdoor lighting is low level with strategically placed fixtures; it is designed to create contrast rather than to reproduce daylight conditions. An important point to watch is that a floodlight not be directed into oncoming traffic—either pedestrian or vehicular.

14.15 Summary

As we have pointed out in this chapter, there is enough latitude in lighting that the design of lighting installations can vary with the degree of design experience and money available. Therefore, getting the best results for the least money requires a tradeoff, i.e., engineering various schemes versus taking the first design. Thus, there is a certain art to lighting design. We have not delved deeply into the influence of colors or into the tricks of the trade. If the reader desires further information, his source of information should be "The Society of Illumination Engineers' Handbook."

Chapter Fifteen

Checking

15.1 Drawing Issue

Before a drawing is issued, it must have the necessary signatures of approval. One of these signatures is placed in an allocated space marked "Checked—Date." A signature in this spot states that this person reviewed the drawing and that first, there are no technical errors, and second, the drawing methods and symbols and the client's wishes have been followed within the scope of recorded information in the job file, and within company policies. There is also a place for an approval signature. It should be emphasized that if a drawing must be issued without check, then it must be issued unsigned as a preliminary, until someone takes the time to check it and sign it as checked. The approval signature authorizes a drawing for issue, but it does not state that a drawing has been checked.

15.2 Conditions of Issue

In most cases, when drawings are being prepared for issue, the pressure is on to meet the deadline. As checking is one of the last routines, the tendency is to rush it through. If this is done, it must involve the calculated risk of allowing errors to pass.

15.3 Drawing Check

Theoretically, a design is completed in rough form and presented to the draftsman for drawing on formal paper. When the draftsman has completed the drawing, it is returned to the designer or sent to a checker. It is then completely checked. After the check, it is returned to the draftsman so that the corrections may be made. When this is complete, it is returned to the checker for *back check*. Back check involves a check to see if all the corrections called for were "picked up" correctly by the draftsman. This, then, should conclude the checking, i.e., check, correction, and back check. In actual practice, this procedure is rarely adhered to for many reasons—but it should be the routine to strive for.

15.4 Color Code Checking

Most engineering offices follow a practice of color coding the check marks and corrections: red, insert; green, delete; yellow, okay. Sometimes the green or the red may change, but yellow seems to be the universally accepted color for okay or all correct. For a drawing to be completely checked, *every* line on the drawing must be colored. This is for the very simple reason that if every line is checked, then it must be either correct (yellow), wrong, delete (green), or changed, added (red). There are no other possibilities except for "unchecked," which is not permissible.

15.5 Checking Philosophy

There are many ways to design a product and achieve the same end result. Consider the different designs of automobiles. The obvious fact exists that a checker will review a drawing and have a strong tendency to change a perfectly good design. *Do not do it*. A checker must be like a court judge. Whatever he may think personally, he must generally suppress. He must stick exactly to the facts, e.g.,

1. Does the drawing meet the specifications?
2. Does the design work?
3. Are the symbols correct?
4. Are sizes and dimensions accurate?
5. Is there a code violation?

In the checking process, there are no "ifs, ands, or buts." A drawing is either right or wrong. Remember that when the drawing is issued, it is stating, If you build it like this, then you will have exactly what you ordered. In fact, a drawing is a logic proposition; any description is a proposition, and, as such, it must have a valence of T or F, that is, true or false.

Once the concrete is poured, it is too late to say "Sorry, but I didn't intend it to be done this way."

Be objective and check only for errors, omissions, and matters of design policy.

If a check is made for cost analysis it should be done on a final drawing that has been checked. It is usually the practice of a contractor to check the final drawing to see if he can make the installation inexpensive and yet adhere essentially to the approved drawing.

15.6 Drafting Technique

A good drawing will be easy on the eyes. At first glance, installed raceways and installed equipment should practically appear in three dimensions, in contrast to background lines.

Notes should be no more than 5 to 6 in. wide. Each note should be separated from the next by double or triple space. Printing should be firm, plain, and obvious.

The eyes get tired when they constantly pass over the same color density. Therefore, try for contrasts. Use the contrasts to accentuate the design portion of the drawing.

Part Two

Simplified Design Mathematics

Chapter Sixteen

Number Manipulation

16.1 The Scalar

A scalar quantity can be considered an undirected quantity. It has no direction and can be fully described by a number. It is a quantity of magnitude only.

16.2 The Equality

The equation $a = b$ is a simple statement, but for all its simplicity the two parallel bars indicating equality must be quite mangled and bent through misuse and misapplication. The statement of equality means that whatever a represents can be replaced with whatever b represents. A very significant point which is too often overlooked is the validity of the converse condition. For an equality to be valid the converse of the equality must also be valid.

Consider the statement "10 lb of apples is equal to 10 lb of oranges." This is a valid statement when we are considering equal weight; however, to the housewife who is baking apple pie, the 10 lb of oranges will not replace the 10 lb of apples. This leads to the conclusion that the equality cannot be used at random; it must be carefully considered and used within the limitations set for an equality. These limitations we can outline as follows:

An equality is an exact statement of fact for a situation of limited conditions, and to be valid the equality must be commutative.

$$(a = b) = (b = a) \quad \text{Commutative law}$$

16.3 The Sum

The operation of addition is a relatively simple procedure when we are dealing with scalar quantities. Addition is both associative and commutative; in other words, the order in which scalars are grouped or the order in which they are added is of no consequence. The sum of a number of scalar quantities is another scalar quantity.

$$a + (b + c) = (a + b) + c \quad \text{Associative law}$$
$$a + b = b + a \quad \text{Commutative law}$$

16.4 The Difference

Subtraction is sometimes called the inverse of addition; however, one limitation that should be obvious precludes the use of the term "inverse." The implied meaning of the word inverse is to "reverse." When dealing with addition we can reverse the process; however, in subtraction we cannot do this. The commutative law is not valid in the process of finding the difference. One exception exists; this is when $a = b$; then the difference $a - b = b - a$. We can define subtraction in the following manner:

The difference between two scalar quantities is another scalar quantity. When a quantity b is subtracted from a quantity a the result is a third quantity c. When quantity c is added to quantity b, the result will equal quantity a.

$$a - b = c \quad \text{then} \quad c + b = a$$
$$a - b \neq b - a \quad \text{except when} \quad a = b$$

16.5 The Product

Multiplication can be thought of as extended addition. It is commutative, associative, and distributive. Considering these operations individually, we will begin with the extended addition:

1. $ab = a + a + a + a + a + \cdots$ b number of times   Extended addition
2. $ab = ba$ Commutative law
3. $a(bc) = (ab)c$ Associative law
4. $a(bc) = ab + ac$ Distributive law

16.6 Division

Division is a rather "sneaky" operation in that the definition of division is stated by a method of multiplication. In division we have a *dividend*, a *divisor*, and a *quotient*. If we show this in conventional form we have

$$\text{Quotient} = \frac{\text{dividend}}{\text{divisor}}$$

Thus the dividend is a quantity numerically equal to the product of the quotient and the divisor. If we state this in conventional form we have

$$\text{Dividend} = \text{quotient} \times \text{divisor}$$

Considering any number, if

$$\frac{a}{b} = c$$

then a must equal bc, or

$$a = bc$$

With this definition we encounter an extremely important point. This is the condition of *division by zero*. When the condition occurs where the divisor is equal to zero we have a situation that is meaningless or "undefined." If we consider

$$\frac{a}{0} = c$$

for this to be valid according to our definition we must have a product:

$$a = 0 \times c$$

If we substitute numerical values such as 5 and 7 respectively for a and c we have

$$5 = 0 \times 7$$
$$5 = 0$$

This is obviously wrong. When we reexamine our definition of equality we see that this is incorrect. In fact we can make c stand for any number we like and we will still get the same answer; that is, $5 = 0$ or $n = 0$. If we allowed this to happen then mathematics would become chaotic; therefore, to prevent this we will just make the statement, *Division by zero is not permissible.* A word of caution here. The fact that the divisor is equal to zero is not always obvious. Consider the equation

$$\frac{P(q + r)}{a - bc}$$

where $a = 12$
$b = 4$
$c = 3$

Now we have the following:

$$\frac{p(q + r)}{12 - (4 \times 3)} = \frac{p(q + r)}{0}$$

There is no point in proceeding any further with this problem because any answer we would get would be meaningless.

16.7 Unknown Quantities

So far we have considered equality, addition, subtraction, multiplication, and division of scalar quantities. Now we will consider the manipulation of scalar quantities within the scope of these operations. First, let us consider what we are trying to accomplish when we work through a problem. In every case we are trying to arrive at a quantity or value which will equal another set of quantities or values. The quantities we are dealing with may or may not be known quantities, and we are trying to solve for one or more unknown quantities. In the majority of cases only one unknown exists. If the basic operations are so simple how is it that some of the calculations are bothersome? This seems to be due to a number of things, the first being a *lack of understanding of the basic subject*. The second is a *lack of familiarity with the normal operations of transposition and/or manipulation of parameters*. This we will attempt to rectify. We did not cover the operations associated with fractions for the simple reason that a fraction is no more than an operation of division, and as such it can be manipulated very simply.

When we deal with a problem we will use a standard method of writing it down in logical steps. It should become a habitual method; it will lead to a higher percentage of error-free manipulations than will the use of shortcuts or memory-retention methods. When using substitutes for parameters or other quantities we will reserve the beginning of the alphabet (that is, a, b, c, etc.) for known values; for unknown values we will use the last part of the alphabet (that is, x, y, z). On the assumption that we will always be trying to find at least one unknown, we will begin with an equality. This means that we ask ourselves the question

$$y = \text{what?}$$

Assuming that we have the ability to locate the necessary information or known values, we can set up the equation

$$y = a$$

where a is the equivalent of all the known values. If a were equal to $b + c$, then our equation could be (by substitution)

$$y = b + c$$

Now if we go one step further and state that c is equal to d/e, we have the following:

$$y = b + \left(\frac{d}{e}\right)$$

When manipulating quantities, it is a good idea to use signs of aggregation around products and quotients when they are involved in more complex operations. They are not essential but they are a reminder that products and quotients are transposed as single quantities in addition and subtraction operations.

For the quantity b we will substitute the complex fraction

$$b = \frac{p - (fg)}{h(j/k)}$$

This gives us the following expansion of our original equality:

$$y = \frac{p - (fg)}{h(j/k)} + \left(\frac{d}{e}\right)$$

Now we have a problem that may seem a little formidable; however it is a simple matter to substitute all the values and complete the portions within the three sets of parentheses first. For convenience we can temporarily replace a complex part of an equation with a single sign or letter; we can then work through the equation and solve the complex part at our convenience. We will replace all the quantities within the parentheses with the letter n; this letter will represent the actual number that would have been in the aggregators. To distinguish the various n's we will use subscript notation. This means that we identify one from the other by placing a number at the bottom right corner, for example, n_1, n_2, etc. By substitution in our last equation we will obtain

$$y = \left(\frac{p - n_1}{n_3}\right) + n_2$$

where $n_1 = fg$, $n_2 = d/e$, and $n_3 = h(j/k)$.

To complete the problem, we substitute in the parentheses first. This gives another number n_4, and by substituting in the preceding equation we get

$$y = n_4 + n_2$$

We can also convert n_4 into two fractions by dividing both p and n_1 by n_3, giving

$$n_4 = \frac{p}{n_3} - \frac{n_1}{n_3} \quad \text{and} \quad y = \frac{p}{n_3} - \frac{n_1}{n_3} + n_2$$

Now we will review our original expanded equation:

$$y = \frac{p - (fg)}{h(j/k)} + \left(\frac{d}{e}\right)$$

This time we will assume that due to information available we know the value of y, but we do not know the value of e. Now what is the procedure to follow? The first step is to try to set up the equality. That is, the unknown is equal to what? In this case the unknown is e; therefore we must isolate the e and set it equal to everything else in the equation. To do this we can use a manipulation known as *cross multiplication*. We will briefly introduce cross multiplication and then return to the problem.

16.8 Cross Multiplication

Cross multiplication is tremendously useful, and it is also simple to use and easy to understand. It is essentially the operation of dividing or multiplying both sides of an equation by the same number.

Where we have an equality of quotients we have a condition as shown:

$$\frac{a}{b} = \frac{c}{d}$$

It could be more involved with an equality containing numerous quantities; for example,

$$\frac{ab}{cd} = \frac{ef}{gh}$$

When an equation appears in this form we can manipulate the quantities from one side of the equals sign to the other by transferring the quantities diagonally:

$$\frac{ab}{cd} = \frac{ef}{gh}$$

If we wish to solve for a we isolate it by moving all the other quantities diagonally to the other side of the equals sign; this gives

$$\frac{ab}{} = \frac{cdef}{gh} \quad \text{(first move)}$$

$$\frac{a}{} = \frac{cdef}{bgh} \quad \text{(second move)}$$

Now that we have isolated the a, it becomes a simple problem. We must remember that the product is commutative; therefore the order in which the values are transferred does not matter. Make note that the previous example consists of products only. The simplicity of this operation suggests that any equation should be transformed into products if possible. This can be done by the use of aggregators and breaking a complex problem into small sections and substituting a sign or letter temporarily until all manipulations are complete. An example of this will occur in our next problem.

We will now return to our original expanded equation:

$$y = \frac{p - (fg)}{h(j/k)} + \left(\frac{d}{e}\right)$$

We now assume that we know the value of y but that the value of e is unknown. As previously stated, we will use a standard method for working all our calculations. We will now give an example of this procedure. Any other calculations will also be performed in this manner. All calculations will be carried out by numbered lines. We will indicate first the number of the line; then we will give the equation; next we will describe the manipulation that was carried out. This enables us to specifically refer to any part of the solving process. For instance, we could be working on line 9, and we would state "replace in line 3." This would mean that we had changed a part of line 3; by referring to line 3 we can see the change. Now consider the procedure for solving for e in the next example.

194   Simplified Design Mathematics

1. $y = \dfrac{p - (fg)}{h(j/k)} + \left(\dfrac{d}{e}\right)$ Original equation

2. $y = \dfrac{p - n_1 + n_2}{n_3}$ Substitute n_1, n_2, n_3 in line 1

3. $yn_3 = p - n_1 + n_2$ Cross-multiply
4. $n_4 = p - n_1 + n_2$ Substitute n_4 for yn_3
5. $n_4 - p + n_1 = n_2$ Transfer p and n_1
6. $n_4 - p + n_1 = \dfrac{d}{e}$ Replace n_2 in line 5

7. $e = \dfrac{d}{(n_4 - p + n_1)}$ Cross-multiply

8. $e = \dfrac{d}{yh(j/k) - p + (fg)}$ Replace n_4 and n_1 in line 7

Note that in line 5 we transferred p, which is automatically considered plus if no sign is shown; at the same time we moved $(-n_1)$. To transfer plus and minus quantities we simply change the signs and move the quantity to the other side of the equals sign (but *not* diagonally as in cross multiplication). In line 4 we simplified the problem by substituting n_4 for the two quantities. In line 7 we exchange the e and $(n_4 - p + n_1)$ by cross multiplying. In line 8 we replaced all the quantities that had been represented by substitutes. When cross multiplying, take care not to separate a fractional quantity, e.g.,

1. $y = \dfrac{d/e}{h}$

2. $\dfrac{y}{d} = \dfrac{e}{h}$ This is incorrect.

The use of aggregators should be encouraged to prevent accidental separation of quotients. This would prevent the type of error that occurred in the previous example. To isolate the e in the previous problem we proceed as follows:

1. $y = \dfrac{(d/e)}{h}$ Original equation

2. $yh = \dfrac{d}{e}$ Cross-multiply

3. $\dfrac{yh}{d/e} = \dfrac{d/e}{d/e}$ Divide both sides by d/e

4. $\dfrac{yhe}{d} = 1$ Divide both sides

5. $e = \dfrac{d}{yh}$ Cross-multiply

You will notice that we added an extra step in lines 3 and 4 which was actually unnecessary, but it shows how cross multiplying shortens the procedure. We could have gone from line 2 directly to line 5 by cross multiplying.

Having covered the operation of isolating an unknown quantity, we will briefly work through exponents and radicals with a review of their application.

16.9 The Exponent

The *exponent* is an extended product. In the expression a^n, a is the known quantity and n is the exponent on the number indicating the power a is raised to. An exponent is a shorthand way of writing a multiplied by itself n number of times, e.g.,

$$a \times a \times a \times a \times \cdots \quad n \text{ number of times}$$

The n is called the power; it may be integral or fractional. We would call a^n "the nth power of a." An example of this follows:

1. $a = 2$ and $n = 4$
2. $a^n = 2^4$
3. $2^4 = 2 \times 2 \times 2 \times 2$
4. $2^4 = 16$

When multiplying exponents with the same base (the base is the known quantity) we simply add the exponents. For example:

1. $a^2 a^3 = a^{2+3}$
2. $a^2 a^3 = a^5$

We can also add negative powers in the same manner; e.g.,

1. $a^3 a^{-2} = a^{3+(-2)}$
2. $a^3 a^{-2} = a^1$
3. $a^1 = a$
4. $a^3 a^{-2} = a$

Here we raise an interesting point: What is the value of a^0? Again we introduce the all-important zero. We will make the statement that any number raised to the zero power is equal to 1. This may not be obvious at first; however, we will prove this after considering manipulations above and below the line. In a compound expression which is represented as a fraction the quantities below the line may be placed above the line by changing the sign of the power; quantities above the line may be placed below it by following the same procedure.

1. $\dfrac{a^2 b^3}{c^2 d^{-3}} = \dfrac{a^2 b^3 d^3}{c^2}$

2. $\dfrac{a^2 b^3 d^3}{c^2} = a^2 b^3 d^3 c^{-2}$

As the bases are all different this is as far as we can progress with this example. We can see that by changing the sign of the power we can obtain some flexibility in manipulation; we are able to change the problem into multiplication operation if we so desire. Now that we are familiar with this manipulation we can consider the proof of a^0 or n^0 equaling 1.

1. $n^0 = 1$ Original equation
2. $\dfrac{a}{a} = 1$ Equals divided by equals
3. $\dfrac{a^n}{a^n} = 1$ Equals divided by equals
4. $\dfrac{a^n}{a^n} = a^n a^{-n}$ Replacement in line 3 and changing division to product
5. $a^n a^{-n} = 1$ Separation of equality in lines 3 and 4
6. $a^n a^{-n} = a^{n-n}$ Law of exponents
7. $n - n = 0$ Law of subtraction
8. $a^{n-n} = a^0$ Substitution in line 6 from line 7
9. $a^0 = 1$ From lines 5 and 8
10. $a^0 = n^0$ From lines 1 and 9
11. $n^0 = 1$ (except when $n = 0$) from lines 1 and 10

It is now evident that any number raised to the zero power is equal to 1. When a fraction is raised to a power, the fact that the whole fraction is raised to that power is shown by the use of aggregators, for example, $(a/b)^n$. The easiest way to manipulate this type of exponent is to remove the aggregators; this allows us to raise both quantities to the power. For example:

$$\left(\dfrac{a}{b}\right)^n = \dfrac{a^n}{b^n}$$

This allows more flexibility in manipulating the quantities; for example, solving for y in the following equation gives

1. $b = \left(\dfrac{a}{y}\right)^n c$ Original equation

2. $b = \dfrac{a^n c}{y^n}$ Separation of quantities

3. $y^n = \dfrac{a^n c}{b}$ Cross multiplication

4. $y = \left(\dfrac{a^n c}{b}\right)^{1/n}$ Fractional exponents (see Sec. 16.10)

All the preceding steps should be familiar except possibly the one in line 4; here we introduced a *fractional exponent*. The fractional exponent is actually another way of representing a *radical*. This will be covered in the next section when radicals are introduced.

To complete the section on exponents we will list some *laws of exponents;* these can be used for reference.

1. $y^a y^b = y^{a+b}$
2. $(y^a)^b = y^{ab}$ $(y^a)(y^a) \cdots$ b times
3. $(yz)^a = y^a z^a$ $(yz)(yz) \cdots$ a times
4. $(y/z)^a = y^a/z^a$ Where $z \neq 0$
5a. $(y^a/y^b) = y^{a-b}$ If $a > b$ and $y \neq 0$
5b. $(y^a/y^b) = 1/y^{a-b}$ If $a < b$ and $y \neq 0$
6. $y^{a/b} = y^{(1/b)a}$
7. $y^{(1/b)a} = (y^{1/b})^a$

16.10 The Radical

The radical could be called the inverse of the exponent. The operation of finding the principal "root" of a radical is usually done with the aid of tables, slide rule, or logarithms. When finding the principal root of a radical we are given the instructions in shorthand form. The radical $\sqrt[n]{a}$ informs us that we are to find a number which when multiplied by itself n number of times will be equal to a. The n is termed the *index* or *order* of the radical; the quantity a is called the *radicand*. The index may be integral or fractional. An example will show a close similarity between exponents and radicals.

1. $\sqrt[n]{a} = b$
2. $a = b^n$
3. $a = b \times b \times b \times \cdots$ n number of times
4. $\sqrt[n]{a} = a^{1/n}$

We see that in line 4 we have a fractional exponent, and we can also see that it can be shown as a radical.

Consider the following example:

1. $\sqrt[3]{64} = 4$
2. $64 = 4^3$
3. $64 = 4 \times 4 \times 4$

If the radical is equivalent to a fractional exponent then the rules of operation for a radical should be similar to the rules for exponents. If we follow similar procedures there should be no problems. We will now introduce the laws for radicals. All the numbers are considered to be *real* numbers or positive.

1. $(\sqrt[n]{a})^n = a$
2. $\sqrt[n]{ab} = \sqrt[n]{a} \sqrt[n]{b}$
3. $\sqrt[n]{\dfrac{a}{b}} = \dfrac{\sqrt[n]{a}}{\sqrt[n]{b}}$
4. $\sqrt[m]{\sqrt[n]{a}} = \sqrt[mn]{a}$
5. $\sqrt[n]{a^m} = (\sqrt[n]{a})^m$
6. $(\sqrt[n]{a})^m = a^{m/n}$
7. $\sqrt[n]{0} = 0$

When the index is fractional or nonintegral then the use of logarithms is necessary to determine the principal root. Consider the radical $^{1.6}/a$. This would normally require the use of logarithms to extract the root. As this requirement occurs very infrequently we will not cover the subject of logarithms in this text. Before leaving the subject of radicals we will give a typical example; consider the equation in line 1 and solve for y.

1. $b = \sqrt[n]{\left(\dfrac{a}{y}\right)} c$ Original equation

2. $b = \dfrac{\sqrt[n]{ac}}{\sqrt[n]{y}}$ Separation of radical

3. $\sqrt[n]{y} = \dfrac{\sqrt[n]{ac}}{b}$ Cross multiplication

4. $(\sqrt[n]{y})^n = \left(\dfrac{\sqrt[n]{ac}}{b}\right)^n$ Raising both sides to equal powers

5. $y = \dfrac{ac^n}{b^n}$ Law of radicals and law of exponents

16.11 Powers of Ten

When using numbers of large sizes or numbers with large differences it is sometimes easier to reduce the numbers by use of the *powers of ten*. We are already familiar with exponents, and so the following manipulations should be recognizable. By multiplying a number by 10 we basically move the decimal point, for example, $97 \times 10 = 970$; also, if we multiply 97 by 10×10 we get 9,700, and $97 \times 10 \times 10 \times 10$ gives 97,000. We will recognize that $10 \times 10 \times 10$ is equal to 10^3 and therefore we could show 97,000 as 97×10^3. By observation we see that multiplying by positive powers of ten adds a number of zeros after the number being multiplied; the number of zeros is equal to the power that the ten is raised to. By the same reasoning, if we have a negative power we move the decimal point to the left the same number of spaces as the negative power indicates. Consider the following examples:

1. $97 \times 10^3 = 97,000$
2. $97,000 \times 10^{-3} = 97$
3. $97 \times 10^{-3} = 0.097$
4. $97 \times 10^5 \times 10^{-3} = 97 \times 10^2 = 9,700$
5. $97 \times 10^5 \times 10^{-3} \times 10^{-4} = 0.97$

The following is a typical problem. Calculate the horsepower in the equation in line 1 where T is torque, N is rpm and is equal to 1,650, and hp is horsepower.

1. $\text{hp} = \dfrac{2\pi NT}{33,000}$

2. $\text{hp} = \dfrac{2 \times 3.14 \times 1,650 \times T}{33,000}$

3. $\text{hp} = \dfrac{2 \times 3.14 \times 1.65 \times 10^3 \times 10^{-4} \times T}{3.3}$

4. $\text{hp} = \dfrac{\cancel{2} \times 3.14 \times \cancel{1.65} \times 10^{-1} \times T}{\cancel{3.3}}$

5. $\text{hp} = 3.14 \times 10^{-1} \times T$
6. $\text{hp} = 0.314 \times T \quad$ for 1,650 rpm

We can see that this whole derivation can be accomplished by visual inspection in line 4 but it is not obvious in line 2; also, it is easy to keep track of the decimal point location. Notice should be taken that in line 3 we immediately reduce all the numbers to a value between 1 and 9. This is not absolutely essential, and sometimes it is more convenient to reduce to a number between 10 and 99—this is a matter of personal preference. It is recommended however that once a routine is established, the reader stay with it.

This completes our review of some of the salient points of the manipulation of scalar quantities. In the next chapter we will review the manipulation of a quantity that has both magnitude and direction; this leads to the study of trigonometry. Chapter 18 is the study of manipulation of vector quantities. To take full advantage of the following two chapters the information we have just presented must be fully understood, and the reader should be able to carry out the manipulations instinctively.

Chapter Seventeen

Trigonometry and the Radius Vector

17.1 The Right Triangle

The use of trigonometry in electrical design is essential. It is not necessary to memorize all the various functions. The majority of trigonometry problems encountered in electrical design can be solved by the functions associated with the right triangle and with an understanding of the so-called trigonometric tables.

Based on past experience and in keeping with the approach that we will present only material that is in common use, we will introduce the trigonometric functions known as the sine, cosine, and tangent. These are usually abbreviated sin, cos, and tan.

The study of trigonometry as a whole should be a planned program. Anyone not having reached this level and intending to pursue more advanced electrical engineering studies should definitely follow an accredited course. This section however will supply practical methods and the understanding necessary to comprehend the electrical design problems in this text.

The trigonometric tables are a list of ratios of the sides of a right triangle. We identify the three sides of a right triangle as shown in Fig. 17.1. The hypotenuse can be considered a vector quantity in that it has magnitude and direction when referred to either of the other

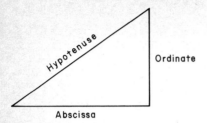

Fig. 17.1 The right triangle.

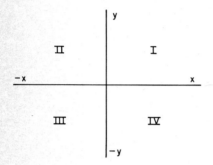

Fig. 17.2 The coordinate axis.

two sides of the triangle. The left corner of the right triangle originates at the junction of two lines at right angles to each other; these lines are known as the *coordinate axis*.

17.2 The Coordinate Axis

The ordinate and the abscissa are the vertical and horizontal axes of a set of lines known as the coordinate axis. The standard representation of the coordinate axis consists of four quarter sections numbered counterclockwise from the top right quarter. The ordinate is the vertical axis and is identified by the letter y above the horizontal axis and $-y$ below the horizontal axis. The horizontal axis is the abscissa and is identified by x on the right side of the vertical axis and $-x$ to the left of the vertical axis. This is shown in Fig. 17.2.

17.3 The Radius Vector

The reference point, the 0° or 360° point, is always to the right of the x axis. A vector is always seen as rotating in a counterclockwise direction. A vector quantity has magnitude and direction; therefore to express a vector quantity a shorthand method is used which indicates its magnitude and direction. Consider a force of 10 units at an angle of 30°. This would be written 10/30°, the magnitude followed by the angle. This is shown in Fig. 17.3, where we have a radius vector 10 units long and 30° from the reference point (the *x* axis). The radius vector could be the resultant of two or more other vectors; this is illustrated in Fig. 17.4.

17.4 Trigonometric Functions

Obtaining the values of two vectors (at right angles to each other) which would result in a single radius vector (as shown in Fig. 17.4) requires the use of trigonometric functions. These functions are merely ratios of one side to another. There are three functions and three inverse functions, but we will consider only the three initial functions; these are the cosine, the sine, and the tangent. To represent the numerical value of the function we use the notation cos θ, sin θ, and tan θ. Cos θ, for example, means "the cosine of the angle θ degrees" where θ can be any number of degrees. These values are obtained from the following equations:

$$\cos \theta = \frac{\text{abscissa}}{\text{hypotenuse}}$$

$$\sin \theta = \frac{\text{ordinate}}{\text{hypotenuse}}$$

$$\tan \theta = \frac{\text{ordinate}}{\text{abscissa}}$$

Using the above relationships we can now determine the value of the two vectors having a resultant of 10/30°. First we will find the value of the horizontal vector (or the abscissa). To do this we use the first equation and find cos θ. Using our standard form for solving a problem we begin with the original equation in line 1. Then, substituting any

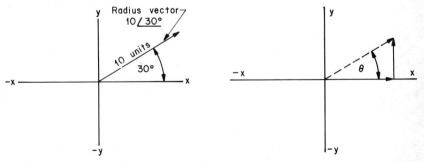

Fig. 17.3 Radius vector 10/30°.   **Fig. 17.4** Two components of a vector.

known values and using our mathematical manipulations, it is relatively simple to arrive at the answer; e.g.,

1. $\cos\theta = \dfrac{\text{abscissa}}{\text{hypotenuse}}$ Original equation

2. $0.866 = \dfrac{\text{abscissa}}{10}$ Substitution of $\cos 30°$ from trigonometric tables

3. $0.866 \times 10 = \text{abscissa}$ Cross multiplication
4. $\text{Abscissa} = 8.66$ Commutative law

Now we consider the value of the ordinate or vertical vector. Here we will use the second trigonometric function and solve for $\sin\theta$. Again using our standard procedure we have in line 1 the original equation.

1. $\sin\theta = \dfrac{\text{ordinate}}{\text{hypotenuse}}$ Original equation

2. $0.5 = \dfrac{\text{ordinate}}{10}$ Substitution of $\sin 30°$ from trigonometric tables

3. $0.5 \times 10 = \text{ordinate}$ Cross multiplication
4. $\text{Ordinate} = 5.0$ Commutative law

17.5 Pythagorean Theorem

To add two vectors we must obtain the vector sum of the two magnitudes and the vector sum of the two angles. This will result in a single vector. In Sec. 17.4 we considered vector quantities on the x and y axes, 90° apart. When this condition occurs we can add by the following method:

$$a^2 + b^2 = c^2 \quad \text{Pythagorean theorem}$$

where c is the hypotenuse and a and b are the other two sides. Using values from Sec. 17.4, we have $10/\underline{30°}$ equals the vector sum of 8.66 and 5 and 90° apart.

1. $a^2 + b^2 = c^2$ Original equation
2. $8.66^2 + 5^2 = 10^2$ Substituting
3. $75 + 25 = 100$ Adding the squares
4. $\sqrt{100} = 10$ Solving for c

This gives us the magnitude of the new vector. From these relationships it would seem that if we can find the magnitudes of the two sides resulting in a unit vector we should also be able to determine the direction or angle of the unit vector. In our previous problems we found the abscissa, ordinate, and radius vector. We know from our trigonometric relationships that the ordinate divided by the abscissa will give the tangent of the angle. It is then a simple procedure to check the tables and find the angle. To complete the above problem and derive the angle we set up our first equation:

1. $\tan \theta = \dfrac{\text{ordinate}}{\text{abscissa}}$ Original equation

2. $\tan \theta = \dfrac{5}{8.66}$ Substitute

3. $\tan \theta = 0.577$ Find tangent
4. $\tan^{-1} 0.577 = 30°$ Find θ

The -1 in $\tan^{-1}$ is not an exponent; this is a shorthand method of writing "an angle whose tangent is . . . ." The -1 has no numerical value; it may also be used for the sine and cosine, e.g., $\sin^{-1} 0.5$ and $\cos^{-1} 0.866$.

17.6 Common Functions

It is suggested at this point that the reader memorize certain trigonometric functions and their values. This is important and helps one obtain a feel for the problem as it is progressing. By examining the values of 0°, 30°, 45°, 60°, and 90° we will see that a definite relationship between them develops. Memorizing the following six numbers gives access to the values of twelve functions.

1. $0 = \sin 0° = \cos 90° = \tan 0°$
2. $1 = \cos 0° = \sin 90° = \tan 45°$
3. $0.866 = \cos 30° = \sin 60°$
4. $0.707 = \cos 45° = \sin 45°$
5. $0.577 = \tan 30°$
6. $1.73 = \tan 60°$

Knowing the above values is of immense importance when it is necessary to perform calculations on the spot, i.e., in a meeting or out on a construction site.

17.7 Quadrants

The majority of electrical design calculations requiring the use of trigonometry will fall into the "first quadrant" category. The quadrants are shown in Fig. 17.5.

17.8 Angles Greater than 360°

A vector is normally rotated counterclockwise and is not limited to one rotation, i.e., 360°. When an angle exceeds 360° it will obviously have

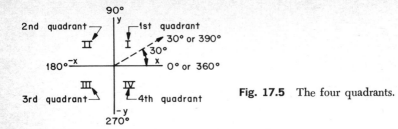

Fig. 17.5 The four quadrants.

to pass through one or more of the quadrants a second time. The value of the trigonometric functions will be equivalent to the values given for 0° to 90°; however, the numbers will be prefixed with a plus or minus sign depending on the quadrant the unit vector is in. Consider an angle of 390°. To find its trigonometric functions we subtract 360° from 390°; this leaves us with 30°. The cosine of 30° is 0.866 and the cosine of 390° is also 0.866. Even though the values are the same the angles are not. This point is very important for a reason that may not seem very obvious. Trigonometry as we use it in electrical design is only two-dimensional. But consider a vertical threaded bolt, and a nut which has been tightened. Instructions may be given to loosen the nut by 390°; this would mean one full turn and a partial turn, and yet if we loosened the nut by 30° we would not be complying with the instructions. Just because the values of the trigonometric functions of an angle equal those of an angle under 360°, do not make the mistake of assuming that the angle itself is necessarily less than 360°.

17.9 Functions of Angles Greater than 90°

Inspecting the trigonometric tables, we find that they stop at 90°. How then do we find the values of the trigonometric functions of an angle that exceeds 90°? Consider the quadrants and functions as shown in Fig. 17.6. By changing quadrants we change the operational sign of the function. The angle of any radius vector is measured from the abscissa, i.e., the horizontal axis.

When a vector is in the first quadrant the procedure is obvious; we have tables to cover 0° to 90°.

When a vector is in the second quadrant the trigonometric function would be the function of 180° minus the angle; e.g., cos 150° would be equivalent to cos (180° − 150°) = cos 30°; from memory we know that this is equal to 0.866. Because the vector is in the second quadrant (and from Fig. 17.6) we see that the cosine is prefixed with a minus sign, for example, —0.866. Writing this in conventional form, we would have

1. cos 150°
2. cos 150° = (—) cos (180° — 150°)
3. cos 150° = (—) cos 30°
4. cos 150° = (—) 0.866   Parentheses are optional

For the sine of 150°, we would have

1. sin 150° = (+) sin 30°
2. sin 150° = (+) 0.5   Note that the sin is plus in the second quadrant

Another example would be a 210° angle. This would be in the third quadrant. Now we are considering an angle measured below the x axis. For this condition we reverse the process and subtract 180° from the angle. Let us find the cosine of the 210° angle.

1. cos 210° = — cos (210° — 180°)
2. cos 210° = — cos 30° (third quadrant)
3. cos 210° = —0.866

In the third quadrant we notice that the cosine and sine are prefixed with minus signs.

In the fourth quadrant we follow a similar procedure except that we subtract the angle from 360°. Find the cosine of 330°.

1. cos 330° = cos (360° — 330°)
2. cos 330° = cos 30°
3. cos 330° = 0.866   In the fourth quadrant the cosine is plus but the sine and tangent are minus

Figure 17.7 should help to clarify the method of determining the functions of any angle; you will notice that the y axis is not used for deter-

	II	I	
+ sin — cos — tan		+ sin + cos + tan	
	III	IV	
— sin — cos + tan		— sin + cos — tan	

Fig. 17.6 Quadrant operational signs.

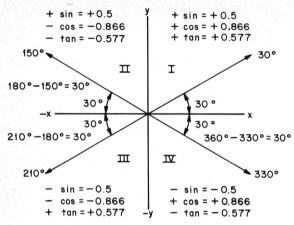

Fig. 17.7 Functions of 30°, 150°, 210°, and 330°.

mining the normal trigonometric functions. Figure 17.7 shows the method of determining four angles; although the angles are different the functions are numerically identical except for the operational prefix of plus or minus. It is essential that the diagram showing the quadrants and the operational prefixes be memorized; after a short period of time, working this type of problem will become routine. When prefixing operational signs it is always helpful to construct the preceding diagram if in doubt.

Chapter Eighteen

Vector Manipulation

18.1 Vector Representation

From the previous chapter we are familiar with the radius vector. The question now arises, How do we handle more than one vector—how do we add, subtract, divide, and multiply vectors without having to draw scale diagrams? The answer to this question is simple; we use manipulations such as those that have already been covered in the previous chapters. The various operations may have different names and may be presented in different form but the actual manipulations are mostly addition, multiplication, division, subtraction, and square roots. Cross multiplication should be routine by this time.

Although the various manipulations are relatively simple, the layout of the problems is usually cumbersome. It requires patience and care to arrive at a successful conclusion. It is impossible to rush through these types of problems and guarantee the results. When working with algebra and arithmetic we sometimes know the approximate answer, and consequently we can recognize if a "slip" has been made along the way. When working with vectors the same familiarity may not be present.

18.2 Location of a Point

Take the familiar x and y axes; then mark a dot or point in one of the quadrants. Now consider the problem of explaining exactly where that point is with respect to the x and y axes. One probably obvious solution is to locate it by using a radius vector with the point at the end of the radius vector; this was covered in the preceding chapter. At this stage we will digress slightly and review briefly the preceding chapters and place them in their exact context because we are now going to tell you to forget some of the previous information. The information was necessary but only for the particular subjects being treated earlier.

The early chapters in Part 2 dealt with the manipulation of numbers; this was a prerequisite to the discussion of trigonometry. In the section on trigonometry we covered angles and how the trigonometric functions are derived. This was for the purpose of understanding how to work with angles. Now we come to the final objective, the manipulation of vectors. Remember that all the preceding information gave you mathematical tools; use the correct tool for the job and *do not try to make the job fit the tools at hand.*

To return to our subject, there are three methods of representing a point on the xy plane. These are *polar, rectangular,* and *cartesian.*[1] None of these methods is recognizable as having been covered in Chap. 17, but a knowledge of trigonometry is required to work with them. In electrical engineering the most commonly used methods are polar and cartesian representation. Each one has advantages and disadvantages; we will consider only the advantageous use of each system. We will introduce a method of vector manipulation that uses the simplest and "least laborious" portion of each system, shifting from one to the other as we see fit.

The aim of this approach is to provide a minimum of complexity and yet to provide adequate information on vector manipulation to enable the electrical designer to solve the problems that he will encounter with vector quantities. The designer should always remember that other persons will probably have to review his calculations. Thus he should try to use a simple approach.

18.3 Polar Form

We shall consider the polar form of representation first. We have a point identified as P and another identified as Q with actual numerical examples. These points are shown in graphic form in Fig. 18.1. Point P can be located by stating the magnitude of r (radius vector) between points O and P and also by stating the angle between the radius vector and the horizontal axis. Point Q is located by a radius vector of 20 units and an angle of 330° from the x axis. In Fig. 18.1, point Q was shown as $Q = (20,330°)$; this is perfectly correct and an acceptable

[1] Cartesian representation is generally identified as *complex notation.*

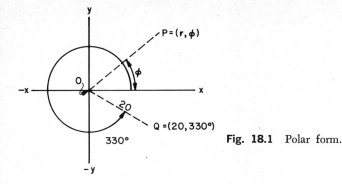

Fig. 18.1 Polar form.

method of writing the location of the point. In electrical engineering we use the same values but the concept changes slightly. Electrical problems are concerned with the length of a radius vector and its location in space rather than with the location of a dot or point. The usual method of writing a vector of 20 units magnitude at 330° is 20/330°. This was explained in Sec. 17.3. It is essential that when we assert something, we realize exactly what we mean, and, furthermore, that we write it in such a way that other people will understand it. To be specific, take point P, for example. Although point P is located by a radius vector and an angle, this does not make it a vector quantity—it is still only a point. The radius vector and angle in this case are reference values only. Alternatively we can have a voltage of 20/330°; this is an electromotive force with an exact location in space with respect to the x axis. The representations then can be used in two ways, i.e., to locate a point or to express a vector quantity. If we recall from trigonometry, we can also express 20/330° as 20/−30°; they are equivalent forms.

18.4 Rectangular Form

When we show a point on a graph, we can locate it in a very simple way. We just say "2 units right and 4 units up"; we then have a specific location. This is the system of rectangular coordinates. The "2 right and 4 up" is expressed (in a shorthand form) as (2,4); x precedes y in the alphabet, and so it precedes it in the representation of x and y coordinates. The same procedure is followed in the other three quadrants. Consider a point in the third quadrant (−2,−4); this means 2 units to the left on the x axis and 4 units down on the y axis. When using rectangular coordinates we will stay with the convention of stating the x coordinate first. We know from trigonometry that if we have the x and y values we can determine the magnitude and angle of the radius vector. Where we used the terms abscissa and ordinate for the x and y axes respectively in trigonometry, we will, in dealing with rectangular coordinates, use only the x and y representation. In Fig. 18.2, polar and rectangular equivalents are given.

18.5 Cartesian Form

A third way of locating a point or specifying a vector quantity is by use of cartesian representation. In general mathematics textbooks this form of identification is referred to as a *complex number;* the two terms are equivalent except that a complex number uses the symbol i instead of the j that is used in the cartesian form. We will employ the notation j and refer to the cartesian form of representation throughout this book. To recap briefly, there are three ways of locating a point, i.e., polar, rectangular, and cartesian. In symbolic form these would be

$$P = r\underline{/\theta} = (x,y) = \pm a \pm jb$$

To explain the construction of a cartesian number we will first reintroduce the rectangular form of a number. This is (x,y). By replacing the x and y values with a trigonometric equivalent and adding an operational sign j we obtain the cartesian form.

From the familiar rectangular form (x,y) we will separate the x and derive its equivalent in terms of the cosine. Next we will take the y and derive its equivalent in terms of the sine.

1. (x,y) — Rectangular coordinates
2. $x^2 + y^2 = r^2$ — Squaring x and y
3. $r = \sqrt{x^2 + y^2}$ — Commutative law and solving for r

From the above we can see that we can obtain the magnitude of the

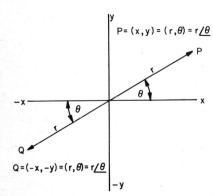

Fig. 18.2 Rectangular and polar representation.

radius vector. Now using trigonometric equations will will replace x in terms of cos θ.

4. $\cos \theta = \dfrac{x}{r}$ Trigonometric function

5. $r \cos \theta = x$ Solving for x, that is, cross multiplication

Now we have x in terms of cos θ and the length of the x component of the radius vector. We will follow a similar procedure with y.

6. $\sin \theta = \dfrac{y}{r}$ Trigonometric function

7. $r \sin \theta = y$ Solving for y, that is, cross multiplication

The x and y components of the radius vector are now represented by $r \cos \theta$ and $r \sin \theta$ respectively. Thus we may assert the following:

8. (x,y) is equivalent to $(r \cos \theta, r \sin \theta)$ Lines 1, 5, 7

If we replaced the r and the functions with numbers we would have only scalar quantities. How then can this method of representation give magnitude and direction? This is accomplished by the use of an operator j. This operator rotates a vector 90° in a counterclockwise direction. Consider line 7 in the preceding analysis. The y which is normally the vertical axis is replaced with $r \sin \theta$ which is a scalar quantity and which could be shown on the x axis. It is however 90° removed from where it should be. To correct this condition we attach the j operator, which locates it in its correct vertical position. In line 9 then we have:

9. $(x,y) = (r \cos \theta, jr \sin \theta)$ Added j operator

From our trigonometry section we know that the functions can be preceded by a plus or minus sign, and so to complete the representation we have

10. $\pm r \cos \theta \pm jr \sin \theta = (x,y)$ Add plus and minus and commutative law

To put the whole of line 10 into abbreviated form we substitute a for $r \cos \theta$ and b for $r \sin \theta$ as in line 11.

11. $\pm a \pm jb = (x,y)$ Substitute in line 10

This, then, is how we obtain the cartesian form of representation for a vector quantity or how we locate a point. This is shown in Fig. 18.3.

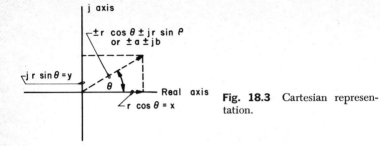

Fig. 18.3 Cartesian representation.

18.6 Polar, Rectangular, and Cartesian Transposition

In the cartesian representation we have a vector quantity shown as a single radius vector but expressed (indirectly) as the x and y components. If we wish to transpose to polar form we must obtain the magnitude and angle; if we have the x and y components then from Chap. 17 we should remember how to obtain the tangent of the angle by dividing the y value by the x value. Consider then the following:

1. $\tan \theta = \dfrac{y}{x}$ Basic equation

2. $\tan \theta = \dfrac{r \sin \theta}{r \cos \theta}$ Replacement in line 1

Remember that the function can be preceded by a plus or a minus sign. Finding the magnitude of the vector is also a familiar procedure. We use the Pythagorean theorem, that is, $c^2 = a^2 + b^2$, where c is the radius vector. One point we must mention here is that the magnitude is derived by the vector addition of a and b; any minus signs would be ignored and replaced with plus signs. For example, the vector $-a - jb$ would have a magnitude of $c = \sqrt{a^2 + b^2}$ but the angle would be $-a/-b$. This will be more obvious as we progress through examples in the next section.

18.7 Operator j

In introducing the operator j formally, we will state that *j is an operator which will rotate a vector 90° in a counterclockwise direction without changing the magnitude of the vector.* The j is not to be confused with a variable; for instance, jb does not mean j multiplied by b. It is simply a shorthand method of saying, "Rotate the vector b 90° to the left," i.e., counterclockwise. The j has a numerical equivalent; this is the imaginary number $\sqrt{-1}$. The definition of imaginary is "not real." If we apply this to $\sqrt{-1}$ we can say that this number is not a real one in the same sense that 1, 2, 3, etc., are. We would never have any reason to derive $\sqrt{-1}$; we also see the impossibility of obtaining $\sqrt{-1}$ when we remember that finding a square root means finding a number that when *multiplied by itself* will give the number under the radical sign. If we consider $\sqrt{1}$ we would have $1 \times 1 = 1$ but we have $\sqrt{-1}$ which would have to be $1 \times (-1) = -1$. We can see then that it is not possible to extract the root, and hence the term *imaginary number*. In our case when we use it in place of j we can use limited mathematical operations. This substitution is used when we are dividing numbers in cartesian form. We will never employ them in this text; instead we will use the simple procedure of switching from cartesian to polar form. This will be outlined as we progress. To continue our discussion of the j operator, consider a vector P. If P is operated on by j we have jP or $\sqrt{-1}\ P$. The vector is rotated 90° counterclockwise. $j^2P = jjP = -1P$ and is rotated 180° counterclockwise. $j^3P = jjjP = j - 1P$ and is rotated 270° counterclockwise. $j^4P = j^2j^2P = (-1)(-1)P = P$; this gives a 360° rotation. In the following examples we will use a vector of 10 units at 30° from the x axis in each of the quadrants (see Fig. 18.4). Changing each polar vector into cartesian form, we have

1. $10/30° = 10 \cos 30° + j\ 10 \sin 30°$ First quadrant
2. $10/30° = 10 \times 0.866 + j\ 10 \times 0.5$
3. $10/30° = 8.66 + j\ 5.0$

1. $10/150° = 10/180° - 150° = 10/30°$ Second quadrant
2. $10/150° = 10 - \cos 30° + j \sin 30° \times 10$
3. $10/150° = 10 \times -0.866 + j\ 0.5 \times 10$
4. $10/150° = -8.66 + j\ 5.0$

1. $10/210° = 10/210° - 180° = 10/30°$ Third quadrant
2. $10/210° = 10 - \cos 30° + j\ 10(- \sin 30°)$
3. $10/210° = 10 \times -0.866 + j\ 10(-0.5)$
4. $10/210° = -8.66 - j\ 5.0$

1. $10/330° = 10/360° - 330° = 10/30°$ Fourth quadrant
2. $10/330° = 10 \cos 30° + j\ 10(- \sin 30°)$
3. $10/330° = 10 \times 0.866 + j\ 10(-0.5)$
4. $10/330° = 8.66 + j\ (-5.0)$
5. $10/330° = 8.66 - j\ 5.0$

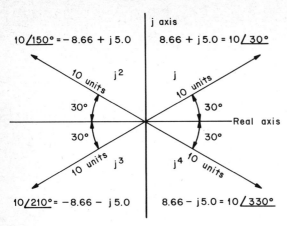

Fig. 18.4 Cartesian and polar equivalents.

18.8 Vector Operations

Now that we have introduced the j operator we can proceed with the application of vector addition, subtraction, multiplication, and division. It is possible to use the cartesian form for all the operations; however, in multiplication and division operations it becomes too cumbersome and consequently conducive to error. Thus we will use the polar form for multiplication and division and the cartesian form for addition and subtraction. In each of the examples below we will use two vectors, that is, $10/30°$ and $10/20°$. We are now familiar with all the following operations. If the previous information was digested the examples should be obvious. We will begin with multiplication; for this we will use the polar form.

1. $10/30°$ multiplied by $20/30°$ *First example*
2. $10 \times 20 / 30° + 30°$
3. $200/60°$

The rule is to *multiply the units and add the angles.*

The second example is division; for this we also use the polar form.

1. $10/30°$ divided by $20/30°$ *Second example*
2. $10/20 / 30° - 30°$
3. $0.5/0°$

The rule is to *divide the units and subtract the angles.*

The third example is addition; for this we use the cartesian form.

1. $10\underline{/30°} + 20\underline{/30°}$ *Third example*
2. $10\underline{/30°} = 10(0.866) + j\,10(0.5) = 8.66 + j\ \ 5.0$
3. $20\underline{/30°} = 20(0.866) + j\,20(0.5) = 17.32 + j\,10.0$
4. Add lines 2 and 3 $\qquad\qquad\qquad\qquad\ \ \overline{25.98 + j\,15.0}$

When you are considering the total as in line 4, remember that the j is only an operator and the plus or minus sign preceding it is dependent on the algebraic addition of the numbers following it. For example:

1. $\ \ 8.66 + j\ \ 5.0$
2. $17.32 - j\,10.0$
3. $\overline{25.98 - j\ \ 5.0}\quad$ Total

The rule here is that *the sign preceding the j is dependent on the algebraic sum of the numbers following it.*

The fourth example is subtraction of vectors; here again we use the cartesian form.

1. $10\underline{/30°} - 20\underline{/30°}$ *Fourth example*
2. $\quad(8.66 + j\ \ 5.0) - (17.32 + j\,10)$
3. $\quad(8.66 + j\ \ 5.0) + (-17.32 - j\,10)\qquad$ See Note
4. $\quad\ \ 8.66 + j\ \ 5.0$ $\qquad\qquad\qquad$ First part of line 3
5. $-17.32 - j\,10.0$ $\qquad\qquad\qquad\ $ Second part of line 3
6. $\overline{-\ \ 8.66 - j\ \ 5.0}\quad$ Total

NOTE: *For the algebraic subtraction the only rule to remember is,* Change the sign of the subtrahend and add; *this we did in line 3.* In the fourth example we note immediately that the two minus signs in line 6 indicate a vector in the third quadrant. The vector addition of the numbers, that is $\sqrt{a^2 + b^2}$, will give 10 units, and the division of $-5.0/-8.66 = 0.577$ and $\tan^{-1} 0.577 = 30°$, but the operational signs are both minus. This indicates 30° in the third quadrant or $30° + 180° = 210°$ or $10\underline{/210°}$ resultant.

18.9 Vector Manipulation

If we wish to multiply a vector in cartesian form by a polar vector we simply change the cartesian form to polar form and proceed as usual. Consider the following vectors: $25.98 + j\,15$ and $10\underline{/30°}$. We multiply these two vectors as follows:

1. $(25.98 + j\,15) \times 10\underline{/30°}$ Original problem
2. $\sqrt{25.98^2 + 15^2} = 30$ Separation of product in line 1
3. $\tan\theta\;15/25.98 = 0.577$ Finding $\tan\theta$
4. $\tan^{-1} 0.577 = 30°$ Finding angle
5. $30\underline{/30°} \times 10\underline{/30°}$ Substitution of lines 2 and 4 in line 1
6. $30 \times 10\underline{/30° + 30°}$ Multiplication
7. $300\underline{/60°}$ Resultant

NOTE: *Lines 2, 3, and 4 were all devoted to changing the cartesian form to polar form. For another example, consider the division of $8.66 + j\,5.0$ by $17.32 + j\,10$.*

1. $\dfrac{8.66 + j\,5.0}{17.32 + j\,10}$ Original problem
2. $8.66 + j\,5.0 = 10\underline{/30°}$ Separation in line 1
3. $17.32 + j\,10 = 20\underline{/30°}$ Separation in line 1
4. $10/20\underline{/30° - 30°}$ Division
5. $0.5\underline{/0°}$ Resultant

The intermediate steps in lines 2 and 3, i.e., changing from cartesian to polar form, have been omitted on the assumption that they are obvious by now. Now, to recap we will give one final example, which will point out again the significance of the operational signs in the cartesian form of vector representation. We note that in line 2 of the preceding example $8.66 + j\,5.0 = 10\underline{/30°}$. We will consider the vector $-8.66 - j\,5.0$; this happens to be equal to $10\underline{/210°}$, and to show this we have

1. $-8.66 - j\,5.0$ Original vector
2. $\sqrt{8.66^2 + 5^2} = 10$ Vector addition
3. $\dfrac{-5}{-8.66} = 0.577$ Finding tangent
4. $\tan^{-1} 0.577 = 30°$ Find angle
5. Continued after Note

NOTE: *In line 2 we ignore the minus signs and use a plus for the vector addition; in line 3 we use the two minus signs to determine the trigonometric function. This does not give us the information to determine the angle because $5/8.66$ also gives 0.577. It may be a good idea at this point to review Sec. 17.9, "Functions of Angles Greater Than 90°." In line 4 we have the value of the trigonometric function; from* this we can determine the angle, but only in the range of 0 to 90°. Next we must consider the two minus signs; the first one (in −8.66) is associated with the sine. We now ask in what quadrant the cosine and sine are negative. By reviewing the four quadrants we find that it is the third. In line 5 then, we have

5. $30° + 180° = 210°$ Third quadrant
6. $10\underline{/210°}$ Resultant as in lines 2 and 5

This concludes our review of vector representation and manipulation. These subjects should be fully understood, so that when you are dealing with electrical problems, you can concentrate on the electrical portion of the problem instead of on the mathematical portion.

Chapter Nineteen

Mathematical Logic

19.1 Principles of Logic

Some dictionaries define "logic" as the "science of correct reasoning." The word "science" in this definition implies an exact form of reasoning. This means that once the principles are established, numerous individuals following those principles will derive the same answer to the same problem. This assumes that no careless errors are made in the manipulations. This also takes logic out of the category of an "art." Art requires "talent" for achievement; in logic this is not the case. If the rules are learned and are not violated, then the derivation of the answer is a foregone conclusion.

In this chapter on mathematical logic we will use the same approach that we maintained throughout the previous mathematics sections. We will present only enough information to ensure that the correct answer can be obtained. We will not introduce material which may be of interest but which, in actual fact, is unnecessary in this text. We will introduce the *implication* in our discussion; this may seem to be in violation of the preceding statement but experience shows that it can be extremely useful. In fact, every individual follows the reasoning of the implication; unfortunately sometimes the analysis of the reasoning violates the rules of logic.

In the engineering profession a system of logic is used for the design of control circuits. These can be electrical, mechanical, pneumatic, hydraulic, etc. The particular system used by engineers incorporates portions of the classical system of logic. This engineering logic is usually called *Boolean algebra,* named after George Boole, a nineteenth-century British mathematician and logician. It is entirely adequate for control circuits but does not provide the tools for a complete logic analysis of any general situation. We will approach the study of logic along the classical lines; the laws and axioms are the same for Boolean algebra, but with the "implication" we will have a broader understanding. Although we will consider *general* problems in logic it is also easy to consider electrical problems using the classical approach. The converse is not necessarily true; electrical logic may not solve some of the general problems.

For the student who wishes to utilize the electronic or electrical Boolean algebra it is only necessary to obtain a text written for the profession. If these lessons have been learned thoroughly then the new text will hold no surprises other than different symbols and signs. Even these will vary from book to book, which is another reason why we will stay with the classical approach.

19.2 The Proposition

A *proposition* is a description. A proposition may be "asserted" if it has a "valence" of T. It may not be asserted if it has a valence of F. To assert something or to make an assertion means to make a positive statement. A statement may be true or false; if it is true it may be written down or spoken. If it is false it may not be written down or spoken unless identified as false. The valence of a proposition is its "truth or falsity." If it is true we use the sign T. If it is false we use the sign F. A proposition is usually in the form of an equality but all equalities are not necessarily propositions. Consider the following example:

1. $4 = 2 + 2$

The above statement would seem to be true but it is not a proposition. It may not even be true. Consider a further example:

2. $(4 \text{ animals} = 2 \text{ sheep} + 2 \text{ cows}) = T$

The above is a proposition with a valence of T, that is, a true proposition. Consider now a third example.

3. $(4 \text{ cows} = 2 \text{ sheep} + 2 \text{ cows}) = F$

This is a false proposition and therefore has a valence of F. A proposition does not have to be in the form of a mathematical equation. It can just as easily be a grammatical expression such as

$$\text{Division by zero is undefined} = T$$

The term on the left side of the equals sign is the description and that on the right side is the valence; the whole is a proposition. The valence of a proposition is not always obvious. In fact, if it were, there would be no need for the other rules of logic. Finding the valence of a proposition is comparable to finding the answer in an arithmetic problem. In both cases we are given a quantity of information which is manipulated according to the rules. If errors are not made we finally reach a point where the information has been compacted into a form that we can apply. Einstein said that $E = mc^2$; this is not a proposition, and converting it into a proposition required a tremendous amount of detailed research and experimentation. This was done during the development of the atomic bomb, and therefore we can now write $(E = mc^2) = T$. We will see later how the use of logic analysis can help in research and experimentation.

19.3 Negation

Determination of the valence of a proposition is not always simple. It may suit our purpose to change the valence from one form to the other—still maintaining the validity of the proposition. To do this we use the operation known as the *law of negation*. It is a relatively simple manipulation. To change the valence of a proposition we simply negate both sides of the proposition. To do this we use a negation sign called a prime which is shown as ($'$). The application of this sign changes a true valence to a negated false valence; i.e.,

1. $T = F'$
2. $F = T'$
3. $T = T''$
4. $F = F''$

NOTE: *Double primes can be used and later on will be used. The negation sign does not make a valence of T into a false valence; it only allows the use of the F sign. As this is negated it still has the function of a T.* Consider for example the proposition

1. Division by zero is undefined $= T$
2. (Division by zero is undefined)$' = T' = F$
3. Division by zero is NOT undefined $= F$

We can see in line 3 that the negation sign has been replaced with the word NOT. To rephrase line 3 we can state, It is false to say that division by zero is not undefined. To introduce our first logical axioms we have

$$T = F' \quad \text{Axiom}$$

and

$$F = T' \quad \text{Axiom}$$

These are the *laws of negation* and they are sometimes known as the *laws of contradiction*.

19.4 The Logical Sum

Here we will introduce the logical sum. The sign used to represent this is the conventional plus sign (+) which we will recognize from our arithmetic. The sign is used to indicate the grammatical connective OR. The OR is used in the inclusive sense and gives us a choice of *one* or *the other* or *both*. It is always used in this sense; do not try to bend it to fit another context. The other use of the OR is in the *exclusive* sense which is an entirely different operation. It does not use the plus sign and does not belong in this section; to avoid misuse of the logical sum we will state that the exclusive OR covers the condition where the choice is *one or the other but* not *both*, that is, $(a = b)'$.

Returning to the logical sum, consider the following example: $x + y = T$ if and only if x is true or y is true or both x and y are true. If we substitute $(2 + 2 = 4)$ for x and $(3 + 6 = 7)$ for y we have

1. $(2 + 2 = 4) + (3 + 6 = 7) = T$

Replacing each addend with the correct valence gives line 2

2. $T + F = T$

We see that only one part of the sum needs to be true for the proposition to be true. Note that we are not interested in the actual numbers themselves. We are only interested in seeing that the individual addends have a valence of T or F. It is also true that in a logical sum containing any number of addends only one addend need be true for the proposition to be true; e.g.,

$$a + b + c + d + e + f + g = T$$

if and only if at least one addend is true. This gives us the axiom $T + a = T$.

19.5 The Logical Product

The next useful operation is the logical product. For the product sign we use a dot ($\cdot$). The product is shown as $x \cdot y = T$ *if and only if x is true and y is true*. The dot represents the grammatical connective AND. The product would then be stated, x and y equals T. It is not a product in the arithmetic sense, that is, x multiplied by y. The product is not limited to any number of factors, but *if any factor is false then the whole proposition is false*. Consider the following example:

1. $a \cdot b \cdot c \cdot d \cdot e \cdot f \cdot g \cdots = T$
   If a, b, c, d, e, f, g are all true
2. $T \cdot T \cdot T \cdot T \cdot T \cdot T \cdot T \cdots = T$
   Replacement in line 1

If any factor had been false then the proposition would have been false. Consider another example. If factor a is false, we would have

1. $a \cdot b \cdot c \cdot d \cdot e \cdot f \cdot g \cdots = F \cdot T$
2. $F \cdot T \cdot T \cdot T \cdot T \cdot T \cdot T \cdots = F$
   Replacement in line 1

In line 1 we set up the problem, but we do not know whether it is true or false; therefore, we set the valence of the product in the form of a logical product. The logical product indicates that the product contains true factors and false factors. In line 2 we replace the factors with their equivalent valence; the first one, a, is false (as specified in the example). The remaining factors can be true or false; this is immaterial. The fact that at least one factor is false immediately establishes the proposition as false. The logical product then is quite simple and it can be defined by the following axiom:

In a logical product, if any factors are NOT true, then the whole proposition is false

19.6 The Logical Equality

Now we will introduce the equality; that is, $a = b$ *if and only if a is false and b is false or a is true and b is true*. The equals sign is equivalent to the grammatical connective "if and only if." An additional condition is attached to the validity of the equality; this is the converse condition. This is extremely important. *The converse of the equality must be true; otherwise the equality is false or invalid*. If the converse is not true then we could have an "implication" which we will cover later. The line of demarcation between an equality and an implication is whether the condition is reversible and still valid in the way it was before. This is not always obvious; therefore, care should be taken in the stating of an equality. This gives the axiom *An equality is valid if and only if a is true* and *b is true* or *a is false* and *b is false* and *the converse condition is true;* i.e., $(a = b) \cdot (b = a) = T$.

The statement of equality is such that

1. $x \cdot y = z$ Original expression
2. $z = x \cdot y$ Commutative law
3. $(x \cdot y = z) \cdot (z = x \cdot y) = T$ If and only if both factors are true
4. $T \cdot T = T$ Substitution of valence in line 3
5. $T = T$ Law of redundancy

As previously explained, logic can be used for analysis of a grammatical statement as well as of a numerical expression. Consider the following example.

Let x represent "horses."
Let y represent "cows are animals."
Let z represent "they have four legs."

If we establish the equality from line 1 in the previous statement we have

1. $x \cdot y = z$ Original equation
2. (horses) · (cows are animals) = (they have four legs)   Replacement in line 1
3. $T \cdot T = T$ Substitution of valence in line 2
4. $T = T$ Redundancy law

This is seemingly a valid equality except that we do not know if the converse is true. In line 5 we set up the converse condition:

5. $z = x \cdot y$ Commutative law
6. (they have four legs) = (horses) · (cows are animals)   Replacement in line 5
7. $T = T \cdot T$ Substitution of valence in line 6
8. $T = T$ Redundancy law

In line 6 it may not be obvious how we arrived at the valence substitution in line 7. In a logic equality we consider the complete statement. Line 6 then would read *"they have four legs if and only if horses and cows are animals."* In line 8 we see that the converse condition is also true, and therefore the equality is valid.

Now consider what would have happened if we had made y represent "men are animals."

1. $x \cdot y = z$ Original expression
2. (horses) · (men are animals) = (they have four legs)   Replacement in line 1
3. $T \cdot T = T \cdot F$ Substitution of valence in line 2

NOTE: *In line 3 we have substituted the valence of T for each product, i.e., "horses and men are animals." Now we consider the valence of "they have four legs"; this is true of horses but untrue for "men are animals if and only if they have four legs." Thus we have the condition where a statement is both true and false at the same time. This is not really very complicated to solve. We have two products; if we reduce them both by the law of products and the law of redundancy, we have line 4:*

4. $T = F$ and is not assertable   Product law and redundancy law

We can now see in line 4 that the original statement in line 1 is false within the framework of the original conditions, i.e., animals have four legs. The proposition then is invalid; a true condition can never equal a false condition. The question now is, Was our logic at fault? The answer to this is no. The answer in line 4 is invalid, and so the original statement is invalid. This situation is covered in a rule called the *vacuity rule*. This states, *If a sign is not governed by any other rule then by this rule that sign is to be regarded as a false proposition.* In other words, if a situation arises that is not covered by any rules, then it should be regarded as false. Line 4 in the preceding example would be covered by this rule. To prove the vacuity rule and also to rework the example we will separate the product in the preceding line 1.

1. $x \cdot y = z$ Original equation
2. $(x = z) = T$ Separation of product in line 1
3. $T = T$ Replacement in line 2
4. $(y = z)$ Separation of product in line 1

Before we show line 5, which will be the replacement of y,z with the valence as in line 3, we will explain how it is derived. Consider the y; this stands for "men are animals"—is this true or not? With reference to the fact that they must have four legs it is false, but considered in a physiological sense it is true, and so we have two conditions. We can say then that the y is both true *and* false, which is $T \cdot F$. Being a product it is equivalent to F; therefore $y = F$. The z is equivalent to "they have four legs." This is true because this is the condition that we set; i.e., *animals have four legs*. From line 3 we replaced z with T; therefore in line 5 we have

5. $(F = T) = F$ Replacement in line 4
6. $(x = z) \cdot (y = z) = T \cdot F$ Lines 2 and 4
7. $T \cdot F = F$ Lines 3 and 5
8. $F = F$ Law of products

In line 8 of the last example we have proved that the original statement as it was worded was incorrect. For it to be true it should have said "*horses* and *men are animals* if and only if (some) *have four legs*." Introducing the word "some" leads into propositional calculus which (like the number calculus) deals with variable or dynamic quantities. It is not our intention to delve this deeply into logic because, as with ordinary mathematics, we could go on ad infinitum. But it serves the purpose of pointing out that a simple change in a word is like opening Pandora's box if you do not know what you are doing.

As we are still considering the equality we will include one last word of caution. Do not use the equality in the place of an "implication" (this will be covered in the next section); remember that the equality is only valid if the converse condition is true.

19.7 The Implication

The use of the implication is not required in electric circuit design, but someone initially must review the specifications or design criteria. In the design specifications we will usually encounter phrasing such as "If the tank fills beyond the 12-ft mark then the overflow pump must start." This approach of "if this—then this" is much more significant than it seems. We will show that the implication is the governing law for all human and mechanical reactions. It can be applied to anything occurring in the universe. Everything that happens is a result of something that happened previously; this is an inescapable fact. The result of something happening may be different from what we anticipated but this does not mean that what happened was wrong. It means that the conclusion was wrong with respect to the assumption made in the first place. The experts claim that it is impossible to think of "nothing"; therefore when we do anything, we think first. The thought may be so fleeting that we may not be aware of it, but nevertheless, the thought is present. The action (or reaction) resulting from the thought we had is termed the *conclusion*. Whether the conclusion was correct or not is a function of the evaluation of the information after it has passed through the analytical process (or the brain). At some time or other everyone has been burned by something that he thought was cold but which in actual fact was hot. This is an example of the assumption and conclusion. In many cases we may say "It wasn't my fault; he told me it was cold." The unending argument then begins as to "who was right." The study of the implication will provide the tools with which to evaluate this type of condition.

The implication consists of two parts. The first is the *premise* or assumption and the second is the conclusion. There are four basic laws of implication; these cover all the situations that come under the definition of implication. The laws of implication are presented in the form of a proposition and they are as follows:

PROPOSITION

Premise → conclusion = valence

$$T \to T = T \quad \text{Axiom}$$
$$F \to T = T \quad \text{Axiom}$$
$$F \to F = T \quad \text{Axiom}$$
$$T \to F = F \quad \text{Axiom}$$

The arrow which is the sign of implication replaces the grammatical connective THEN. If we have an implication $(x \to y)$, it is read *if x then y*. Alternatively it can be read x *implies* y. The general law of implication is that

$x \to y$ *stands for the sentence that is false* if and only if x *is true* and y *is false*.

When we consider all the possible combinations of this statement we arrive at the four basic laws we have just introduced. First we will consider the premise. The *premise* is an assumption. It may be correct or incorrect, this does not really matter. What is important is the conclusion that is drawn from the premise. The meaning of the word conclusion is obvious, and therefore we will not try to define it. Examining each axiom individually, we will begin with $T \to T = T$. This is straightforward. If we have a true premise and draw a true conclusion then we can say that our logical thinking was correct and the proposition is true. The next axiom, $F \to T = T$, is also relatively simple. Consider the situation where false information is given. If the information is recognized as false then a true *conclusion* can still be drawn. Once again our logical thinking is correct and the proposition is true. This incidentally was used by Archimedes to prove that there is not a largest prime number. His approach was to assume that there is a largest prime number. His conclusion was that if it exists he should be able to prove it; and if he could not prove it, then his assumption was false and no largest prime number exists. We can see then that false information can be just as valuable as true information. The next axiom, $F \to F = T$, is also easy to understand but sometimes hard to accept. This deals with the problem of receiving false information under the impression that it is true. The *conclusion* drawn is false, but this is as a result of not knowing that the *premise* was false. Therefore, the logical thinking was correct based on the information given; the proposition then is also true. It just happens to be the case that the information given was false. This is why bridges collapse, buildings fall down, battles are lost. The technician does his work correctly but the information supplied is incorrect.

The last law of implication, $T \to F = F$, is the only one that is expressed as a false proposition. It should be fairly obvious. It states, If true information is given and a false conclusion is drawn then the logical thinking was wrong. With this last law of implication there is no excuse for error; when true information is given and a false conclusion is drawn then it is clear that somebody made a mistake. To convert from an implication to a sum we use the following:

$$a \to b = a' + b$$

19.8 Modus Ponens

The laws of implication can also be used with a different approach; an example of this is the rule of *modus ponens*. This is from the Latin, and it means *the method of asserting the conclusion as a consequence of asserting the premise*. In other words, if we know that the proposition is true, and if we also know the valence of the premise, then by this rule and by our previous axioms of implication we can determine the valence of the conclusion. Consider the following statement: If the *implication* is true and the *premise* is false then the *conclusion* must be true. If we review the four laws of implication we will see that there is no other alternative. Consider another statement: If the *implication* is false and the premise is true then the *conclusion* must be false. This of course is the fourth law of implication ($T \to F = F$). These last two statements are both examples of the rule of *modus ponens*.

Another point to consider with the implication is the converse condition. The implication and the equality are very close to being equivalent operations; the significant difference between the two is the validity of the converse condition of the proposition. In the implication the converse condition may or may not be valid; this we can see from our axioms of implication. The premise may be true or false and a correct conclusion can still be drawn. With regard to the converse, if we have a correct conclusion and the proposition is true then the premise can be either true or false.

19.9 De Morgan's Law

We have now covered some of the basic operations of mathematical logic. These are the sum, product, equality, negation, and implication operations. Sometimes we will encounter an equation that is "bordered." This means that it is enclosed in aggregators. Aggregators used in logic include parentheses, brackets, and braces. The use of the vinculum as an aggregator is not recommended. The aggregators are used in the following order:

$$\underset{\text{First}}{(\quad)} \underset{\text{Second}}{[(\quad)]} \underset{\text{Third}}{\{[(\quad)]\}}$$

When we encounter a bordered equation that is negated, the first step in solving the problem is to remove the aggregators and the negation sign. For this operation we use De Morgan's law. It is relatively simple to use, but again a word of caution: In a complex or lengthy expression it is easy to become careless.

De Morgan's law can only be applied to a sum or a product. Where the bordered and negated expression is an application or an equality it must be changed to its equivalent sum or product. An example of this will be shown when we discuss the exclusive OR.

Using De Morgan's law, we remove the aggregators and change the connecting sign from a dot to a plus or vice versa and negate the parameter. This gives two conditions:

$$(a + b)' = a' \cdot b' \qquad \text{Axiom}$$
$$(a \cdot b)' = a' + b' \qquad \text{Axiom}$$

This operation, as you can see, is simple; therefore we will not give any further examples. De Morgan's law is used repeatedly in logic analysis, and so it should be memorized, along with the other operations we have covered.

19.10 The Exclusive OR

The *exclusive* OR is not a simple expression like the *inclusive* OR. The exclusive OR gives us the choice of *one* or *the other but* not *both*. The expression for this cannot be considered the same as the sum, i.e., the inclusive OR. It is in fact a combination of *sum* and *product*. The standard equation for the exclusive OR is

$$(a = b)' \qquad \text{Axiom}$$

It will not be obvious to the beginner that we have an OR expression when it is shown in this form. For this reason we will present a categorical proof of this expression. (The others you have been asked to accept without proof.)

The operations we will use in the proof have all been covered in our study; we will use nothing that has not been introduced. If an expression is not immediately recognized a review of the previous work should resolve the question.

If an equality is valid then the converse of the equality should also be valid. We also explained that in an implication the converse may or may not be valid. Consider then how we could outline the requirements for an equality by using the limitations of the implication. Take the equality $a = b$. To express this by means of the implication, we would have to say, a implies b and b must imply a. This meets the condition of the equality. To show this in mathematical form we would have

$$(a = b) = (a \to b) \cdot (b \to a)$$

If we remember the law of products we see that both implications must be true and also that one is the converse of the other. If we negate this equality we now have the beginning of our proof.

1. $(a = b)'$ — Original equation
2. $(a = b)' = [(a \to b) \cdot (b \to a)]'$ — Law of equality
3. $[(a' + b) \cdot (b' + a)]'$ — Separation of equality and standardization implication
4. $(a' + b)' + (b' + a)'$ — De Morgan's law
5. $a'' \cdot b' + b'' \cdot a'$ — De Morgan's law
6. $a \cdot b' + b \cdot a'$ — Negation laws
7. $(a \cdot b' + b \cdot a') = (a = b)'$ — Line 6 dropping line 1

If we analyze what we are actually stating in line 7 we could consider the commutated form and begin with the original expression as follows:

$(a = b)'$ is equivalent to a AND NOT b OR b AND NOT a

You will notice that this expression allows no acceptance of both as in the ordinary sum (inclusive OR). You will also notice that without an understanding of the implication it would not be possible to make up a proof. We could use the argument of going from line 1 directly to line 3, but then the question of how to prove line 3 arises. It must be proved before it can be used in a proof.

One of the requirements of a proof is that it only consist of previously proven theorems and axioms. The term "proven theorem" means that the theorem has been justified but it is not possible to use finite values to indicate a proof. An example of this is the statement of Archimedes that there is no largest prime number. Another theorem that is a much better example is the theorem that there is no largest number. We cannot prove this by giving a largest number and saying, See, this is not it. It still remains a theorem because it cannot be proven. We will therefore use the "proven theorem," which will apply to all theorems that can be shown to be obvious. We could reword our statement on the largest number by stating:

By adding the number 1 to whatever number may be thought of as being the largest number, we have a new largest number, and so on, ad infinitum. This then is the reason for not jumping from the original expression to line 3 in our proof. We must have previously shown that $a' + b$ is equivalent to $a \to b$.

We must point out that the reader has been asked to accept all these axioms and definitions without proofs. This approach in one sense is not good, but it has been necessary to keep this section within the limits of a handy reference work rather than a complete study in logic. A course of study in mathematical logic will satisfy these further requirements.

19.11 Application of Mathematical Logic

Mathematical logic is a tremendous tool for reasoning and analysis; it can be used in virtually any area, including the analysis of systems, switching circuits, hydraulic systems, pneumatic systems, logistics, speech, highways and traffic control, etc. The field of application is limitless. The very small section we have covered is only an introduction to the main and most widely used operations. We have not covered the use of truth tables, Venn diagrams, Veitch diagrams, or the propositional calculus with the variable parameters. This is left to the reader who wishes to become a student of this very interesting branch of mathematics. It vastly increases the self-confidence of an individual when he knows that if he makes a statement he can prove that what he is saying makes logical sense.

SOME IMPORTANT THEOREMS OF MATHEMATICAL LOGIC

1. $a + b = b + a$ Commutative law
2. $a \cdot b = b \cdot a$ Commutative law
3. $a + (b + c) = (a + b) + c$ Associative law
4. $a \cdot (b \cdot c) = (a \cdot b) \cdot c$ Associative law
5. $a \cdot (b + c) = (a \cdot b) + (a \cdot c)$ Distributive law
6. $a + (b \cdot c) = (a + b) \cdot (a + c)$ Distributive law
7. $(a \cdot b)' = a' + b'$ De Morgan's law
8. $(a + b)' = a' \cdot b'$ De Morgan's law
9. $a \cdot (a + b) = a$ Absorption law
10. $a + a \cdot b = a$ Absorption law

11. $a \rightarrow b = a' + b$ Implication law
12. $a \rightarrow b = (a \cdot b')'$ Implication law
13. $F = T'$ Negation law
14. $T = F'$ Negation law
15. $a'' = a$ Double negation law
16. $a + a' = T$ Valence law
17. $a \cdot a' = F$ Valence law
18. $a \cdot T = a$ Valence law
19. $a + F = a$ Valence law
20. $a + T = T$ Valence law
21. $a \cdot F = F$ Valence law
22. $(a = b) = (a \rightarrow b) \cdot (b \rightarrow a)$ Equality law
23. The vacuity rule states:

If a sign is not governed by any other rule, then by this rule that sign is to be regarded as a false proposition.

NOTE: *This last rule essentially holds that there will be occasions when a known set of rules will not cover a particular situation. When this occurs the situation should be regarded as false. We may assume then that if in fact the situation were true rather than false, this would eventually come to light through more information being presented. Through the laws of redundancy the true valence would become evident eventually.*

Appendix

TABLE A.1 AMPACITIES Allowable Current-carrying Capacities (Amperes) of Insulated Copper Conductors

Not More than Three Conductors in Raceway or Direct Burial, Based on 30°C, 86°F Ambient (Condensed from National Electrical Code)

Size, AWG or MCM	Maximum Operating Temperature					
	60°C	75°C	85–90°C	110°C	125°C	200°C
	Types of Insulation					
	RUW	RH, RHW Versatol Super Coronol	Paper V-C (V) TA, TBS, SA, AVB	V-C (AVA) Deltabeston V-C (AVL) Deltabeston	AI AIA Silicone	A AA FEP*
		RUH	SIS, MI, V-C (AVB) Deltabeston			
	T-TW Flamenol	THW THWN Flamenol	FEP, FEPB THHN Flamenol			
		XHHW Vulkene	RHH XHHW Vulkene			FEP*
14	15	15	25	30	30	30
12	20	20	30	35	40	40
10	30	30	40	45	50	55
8	40	45	50	60	65	70
6	55	65	70	80	85	95
4	70	85	90	105	115	120
3	80	100	105	120	130	145
2	95	115	120	135	145	165
1	110	130	140	160	170	190
0	125	150	155	190	200	225
00	145	175	185	215	230	250
000	165	200	210	245	265	285
0000	195	230	235	275	310	340
250	215	255	270	315	335	
300	240	285	300	345	380	
350	260	310	325	390	420	
400	280	335	360	420	450	
500	320	380	405	470	500	
600	355	420	455	525	545	
700	385	460	490	560	600	
750	400	475	500	580	620	
800	410	490	515	600	640	
900	435	520	555			
1000	455	545	585	680	730	

Correction Factors for Room Temperature Over 30°C

C	F						
40	104	0.82	0.88	0.90	0.94	0.95	
45	113	0.71	0.82	0.85	0.90	0.92	
50	122	0.58	0.75	0.80	0.87	0.89	

* Special use only; see NEC Table 310-2(a).

TABLE A.2 CONDUIT FILL
Maximum Number of Conductors in Trade Sizes of Conductor Tubing

(Based on NFPA National Electrical Code, Chap. 9, Table 1)

Type letters	Conductor size AWG, MCM	1/2	3/4	1	1 1/4	1 1/2	2	2 1/2	3	3 1/2	4	4 1/2	5	6
Tw, T, RUH, RUW, XHHW (14 through 8)	14	9	15	25	44	60	99	142						
	12	7	12	19	35	47	78	111	171					
	10	5	9	15	26	36	60	85	131	176				
	8	3	5	8	14	20	33	47	72	97	124			
RHW and RHH (without outer covering), THW	14	6	10	16	29	40	65	93	143	192				
	12	4	8	13	24	32	53	76	117	157				
	10	4	6	11	19	26	43	61	95	127	163			
	8	1	4	6	11	15	25	36	56	75	96	121	152	
TW, T, THW, RUH (6 through 2), RUW (6 through 2), FEPB (6 through 2), RHW and RHH (without outer covering)	6	1	2	4	7	10	16	23	36	48	62	78	97	141
	4	1	1	3	5	7	12	17	27	36	47	58	73	106
	3	1	1	2	4	6	10	15	23	31	40	50	63	91
	2	1	1	2	4	5	9	13	20	27	34	43	54	78
	1		1	1	3	4	6	9	14	19	25	31	39	57
	0		1	1	2	3	5	8	12	16	21	27	33	49
	00		1	1	1	3	5	7	10	14	18	23	29	41
	000		1	1	1	2	4	6	9	12	15	19	24	35
	0000			1	1	1	3	5	7	10	13	16	20	29
	250			1	1	1	2	4	6	8	10	13	16	23
	300			1	1	1	2	3	5	7	9	11	14	20
	350				1	1	1	3	4	6	8	10	12	18
	400				1	1	1	2	4	5	7	9	11	16
	500				1	1	1	1	3	4	6	7	9	14
	600					1	1	1	3	4	5	6	7	11
	700					1	1	1	2	3	4	5	7	10
	750					1	1	1	2	3	4	5	6	9
THWN, THHN, FEP (14 through 2), FEPB (14 through 8)	14	13	24	39	69	94	154							
	12	10	18	29	51	70	114	164						
	10	6	11	18	32	44	73	104	160					
	8	3	6	10	19	26	42	60	93	125	160			
	6	1	4	6	11	15	26	37	57	76	98	125	154	
	4	1	2	4	7	9	16	22	35	47	60	75	94	137
	3	1	1	3	6	8	13	19	29	39	51	64	80	116
	2	1	1	3	5	7	11	16	25	33	43	54	67	97
	1		1	1	3	5	8	12	18	25	32	40	50	72
XHHW (4 through 500 MCM)	0		1	1	3	4	7	10	15	21	27	33	42	61
	00		1	1	2	3	6	8	13	17	22	28	35	51
	000		1	1	1	3	5	7	11	14	18	23	29	42
	0000		1	1	1	2	4	6	9	12	15	19	24	35
	250			1	1	1	3	4	7	10	12	16	20	28
	300			1	1	1	3	4	6	8	11	13	17	24
	350			1	1	1	2	3	5	7	9	12	15	21
	400				1	1	1	3	5	6	8	10	13	19
	500				1	1	1	2	4	5	7	9	11	16
	600				1	1	1	1	3	4	5	7	9	13
	700					1	1	1	3	4	5	6	8	11
	750					1	1	1	2	3	4	6	7	11
XHHW	6	1	3	5	9	13	21	30	47	63	81	102	128	185
	600				1	1	1	1	3	4	5	7	9	13
	700					1	1	1	3	4	5	6	7	11
	750					1	1	1	2	3	4	6	7	10
RHW, RHH (with outer covering)	14	3	6	10	18	25	41	58	90	121	155			
	12	3	5	9	15	21	35	50	77	103	132			
	10	2	4	7	13	18	29	41	64	86	110	138		
	8	1	2	4	8	10	17	25	39	52	67	84	105	152
	6	1	1	2	5	6	11	15	24	32	41	51	64	93
	4	1	1	1	3	5	8	12	18	24	31	39	50	72
	3	1	1	1	3	4	7	10	16	22	28	35	44	63
	2		1	1	3	4	6	9	14	19	24	31	38	56
	1		1	1	1	3	5	7	11	14	18	23	29	42
	0		1	1	1	2	4	6	9	12	16	20	25	37
	00			1	1	1	3	5	8	11	14	18	22	32
	000			1	1	1	3	4	7	9	12	15	19	28
	0000			1	1	1	2	4	6	8	10	13	16	24
	250				1	1	1	3	5	6	8	11	13	19
	300				1	1	1	3	4	5	7	9	11	17
	350				1	1	1	2	4	5	6	8	10	15
	400					1	1	1	3	4	6	7	9	14
	500					1	1	1	3	4	5	6	8	11
	600						1	1	2	3	4	5	6	9
	700						1	1	1	3	3	4	6	8
	750						1	1	1	3	3	4	5	8

TABLE A.3 Properties of Conductors

Size, AWG MCM	Area, cir mils	Concentric-lay stranded conductors		Bare conductors		Dc resistance, ohms/M ft at 25°C 77°F		
		No. wires	Diam. each wire, in.	Diam., in.	Area,* sq. in.	Copper		Aluminum
						Bare cond.	Tinned cond.	
18	1,620	Solid	0.0403	0.0403	0.0013	6.51	6.79	10.7
16	2,580	Solid	0.0508	0.0508	0.0020	4.10	4.26	6.74
14	4,110	Solid	0.0641	0.0641	0.0032	2.57	2.68	4.22
12	6,530	Solid	0.0808	0.0808	0.0051	1.62	1.68	2.66
10	10,380	Solid	0.1019	0.1019	0.0081	1.018	1.06	1.67
8	16,510	Solid	0.1285	0.1285	0.0130	0.6404	0.659	1.05
6	26,240	7	0.0612	0.184	0.027	0.410	0.427	0.674
4	41,740	7	0.0772	0.232	0.042	0.259	0.269	0.424
3	52,620	7	0.0867	0.260	0.053	0.205	0.213	0.336
2	66,360	7	0.0974	0.292	0.067	0.162	0.169	0.266
1	83,690	19	0.0664	0.332	0.087	0.129	0.134	0.211
0	105,600	19	0.0745	0.372	0.109	0.102	0.106	0.168
00	133,100	19	0.0837	0.418	0.137	0.0811	0.0843	0.133
000	167,800	19	0.0940	0.470	0.173	0.0642	0.0668	0.105
0,000	211,600	19	0.1055	0.528	0.219	0.0509	0.0525	0.0836
250	250,000	37	0.0822	0.575	0.260	0.0431	0.0449	0.0708
300	300,000	37	0.0900	0.630	0.312	0.0360	0.0374	0.0590
300	350,000	37	0.0973	0.681	0.364	0.0308	0.0320	0.0505
400	400,000	37	0.1040	0.728	0.416	0.0270	0.0278	0.0442
500	500,000	37	0.1162	0.813	0.519	0.0216	0.0222	0.0354
600	600,000	61	0.0992	0.893	0.626	0.0180	0.0187	0.0295
700	700,000	61	0.1071	0.964	0.730	0.0154	0.0159	0.0253
750	750,000	61	0.1109	0.998	0.782	0.0144	0.0148	0.0236
800	800,000	61	0.1145	1.030	0.833	0.0135	0.0139	0.0221
900	900,000	61	0.1215	1.090	0.933	0.0120	0.0123	0.0197
1,000	1,000,000	61	0.1280	1.150	1.039	0.0108	0.0111	0.0177
1,250	1,250,000	91	0.1172	1.289	1.305	0.00863	0.00888	0.0142
1,500	1,500,000	91	0.1284	1.410	1.561	0.00719	0.00740	0.0118
1,750	1,750,000	127	0.1174	1.526	1.829	0.00616	0.00634	0.0101
2,000	2,000,000	127	0.1255	1.630	2.087	0.00539	0.00555	0.00885

* Area given is that of a circle having a diameter equal to the overall diameter of a stranded conductor.

The values given in the table are those given in Handbook 100 of the National Bureau of Standards except that those shown in the eighth column are those given in Specification B33 of the American Society for Testing and Materials, and those shown in the ninth column are those given in Standard No. S-19-81 of the Insulated Power Cable Engineers Association and Standard No. WC3-1964 of the National Electrical Manufacturers Association.

The resistance values given in the last three columns are applicable only to direct current. When conductors larger than No. 4/0 are used with alternating current, the multiplying factors in Table 9, Chapter 9, National Electrical Code, should be used to compensate for skin effect.

TABLE A.4 Circuit-breaker Types

1. THERMAL MAGNETIC—INVERSE TIME/CURRENT
   a. Protects against sustained overloads thermally.
   b. Protects against short circuit magnetically.
   USE AS STANDARD
2. MAGNETIC TRIP ONLY (INSTANTANEOUS)
   a. Protects against short circuit *only*.
   USE IF REQUESTED BY CLIENT
3. NON-AUTOMATIC TRIP
   a. Does not protect against any type of overload or fault.
   b. Used as an isolating and manual switch.
   USE AS REQUIRED
4. HIGH-BREAK* BREAKERS
   a. Same as thermal magnetic in size and characteristics.
   b. Has higher interrupting rating.
   USE AS REQUIRED
5. ADJUSTABLE MAGNETIC TRIP
   a. Provides high, intermediate, and low adjustment for short-circuit protection.
   USE AS REQUIRED
6. INTERCHANGEABLE AND FIXED TRIPS
   a. Some breakers have interchangeable trips. Check manufacturer's catalog.
   USE WHERE POSSIBLE

* General Electric trade name.

TABLE A.5 Circuit-breaker Application Ratings
Quick Selection Guide

Circuit-breaker type		Cont amp rating	No. poles	Volts		Rms	Interrupting ratings based on NEMA test procedure							
							Ac rating, volts						Dc rating, volts	
				Ac	Dc		120	120/240	240	277	480	600	125	250
E100	TE	10-100	1	120	125	Sym Asym	10,000 10,000						5,000	
	TE	10-100	2, 3	240	250	Sym Asym								5,000
	TEF	10-100	1	277	125	Sym Asym				14,000 15,000			10,000	
	TEF	10-100	2	480	250	Sym Asym					14,000 15,000			10,000
	TED	15-100	3	480		Sym Asym			18,000 20,000		14,000 15,000			
	TED	15-100	3	600		Sym Asym			18,000 20,000		14,000 15,000	14,000 15,000		
	THEF	15-30	1	277	125	Sym Asym				65,000 75,000			20,000	
	THEF	15-100	2	600	250	Sym Asym			65,000 75,000		25,000 30,000	18,000 20,000		20,000
	THED	15-100	3	600		Sym Asym			65,000 75,000		25,000 30,000	18,000 20,000		
F225	TFJ, TFK	70-225	2	600	250	Sym Asym			25,000 30,000		22,000 25,000	22,000 25,000		10,000
	TFJ, TFK	70-225	3	600		Sym Asym			25,000 30,000		22,000 25,000	22,000 25,000		

	Frame	Amp Range	Poles	Volts		Type	Interrupting Rating (Amperes)				
	THFK ①	70–225	2	600	250	Sym	42,000	25,000	25,000	22,000		20,000
		70–225	3	600		Asym	50,000	30,000	30,000	25,000		
J600	TJJ, TJK4	125–400	2	600	250	Sym	42,000	25,000	25,000	22,000		10,000
		125–400	3	600		Asym	50,000	30,000	30,000	25,000		
	TJK6	250–600	2	600	250	Sym	42,000	30,000	30,000	22,000		10,000
		250–600	3	600		Asym	50,000	35,000	35,000	25,000		
	THJK	125–400	2	600	250	Sym	42,000	30,000	30,000	22,000		20,000
		125–400	3	600		Asym	50,000	35,000	35,000	25,000		
K1200	TKM8	300–800	2	600	250	Sym	65,000	35,000	35,000	25,000		10,000
		300–800	3	600		Asym	75,000	40,000	40,000	30,000		
	TKM12	300–800	2	600	250	Sym	65,000	35,000	35,000	25,000		
		300–800	3	600		Asym	75,000	40,000	40,000	30,000		
	THKM8	600–1200	2, 3	600		Sym	42,000	30,000	30,000	22,000		20,000
		300–800	2	600	250	Asym	50,000	35,000	35,000	25,000		
	THKM12	300–800	3	600		Sym	65,000	35,000	35,000	25,000		
		600–1200	2, 3	600		Asym	75,000	40,000	40,000	30,000		

TABLE A.6 Motor-branch-circuit Data—460-volt Three-phase AC

Horse-power	Full load amp	Starter NEMA size	Circuit breaker			Power control		Three control	Three power
			Frame size	Trip size	Fuse size				
½	1.0	1	100	15	15	1*		No. 12	No. 12
¾	1.4	1	100	15	15	1*		No. 12	No. 12
1	1.8	1	100	15	15	1*		No. 12	No. 12
1½	2.6	1	100	15	15	1*		No. 12	No. 12
2	3.4	1	100	15	15	1*		No. 12	No. 12
3	4.8	1	100	15	15	1*		No. 12	No. 12
5	7.6	1	100	15	20	1*		No. 12	No. 12
7½	11.0	1	100	20	30	1*		No. 12	No. 12
10	14.0	1	100	30	40	1*		No. 12	No. 12
15	21	2	100	40	60	1		No. 12	No. 10
20	27	2	100	50	70	1		No. 12	No. 8
25	34	2	100	70	90	1		No. 12	No. 8
30	40	3	100	70	100	1¼		No. 12	No. 6
40	52	3	100	100	150	1½		No. 12	No. 4
50	65	3	225	100	175	1¼	¾	No. 12	No. 2
60	77	4	225	125	200	1¼	¾	No. 12	No. 2
75	96	4	225	150	250	1½	¾	No. 12	No. 1
100	124	4	225	200	350	2	¾	No. 12	No. 2/0
125	156	5	400	300	400	2	¾	No. 12	No. 3/0
150	180	5	400	400	450	2½	¾	No. 12	No. 4/0
200	240	5	600	500	600	3	¾	No. 12	350 MCM
Table ↑ 430-150		Motor starter	Short-circuit protection based on Table 430-152			Table 4 conduit size, in.		Table 310-12 75°C RHW or THW	

* ¾ in. allowed if three No. 14 controls are used instead of No. 12.
1. 50 hp and up requires separate conduit for control.
2. Do not use TW unless specified.
3. Notes 8 and 10 to Tables 310-12 to 310-15 and Exception 1—Reference 300-3e exempts derating due to motor control conductors.

TABLE A.9 Temperature Conversion
Degrees Centigrade to Degrees Fahrenheit

C	F	C	F	C	F	C	F	C	F	C	F
−40	−40.0	+5	+41.0	+40	+104.0	+175	+347	+350	+662	+750	+1382
−38	−36.4	6	42.8	41	105.8	180	356	355	671	800	1472
−36	−32.8	7	44.6	42	107.6	185	365	360	680	850	1562
−34	−29.2	8	46.4	43	109.4	190	374	365	689	900	1652
−32	−25.6	9	48.2	44	111.2	195	383	370	698	950	1742
−30	−22.0	10	50.0	45	113.0	200	392	375	707	1000	1832
−28	−18.4	11	51.8	46	114.8	205	401	380	716	1050	1922
−26	−14.8	12	53.6	47	116.6	210	410	385	725	1100	2012
−24	−11.2	13	55.4	48	118.4	215	419	390	734	1150	2102
−22	− 7.6	14	57.2	49	120.2	220	428	395	743	1200	2192
−20	− 4.0	15	59.0	50	122.0	225	437	400	752	1250	2282
−19	− 2.2	16	60.8	55	131.0	230	446	405	761	1300	2372
−18	− 0.4	17	62.6	60	140.0	235	455	410	770	1350	2462
−17	+ 1.4	18	64.4	65	149.0	240	464	415	779	1400	2552
−16	3.2	19	66.2	70	158.0	245	473	420	788	1450	2642
−15	5.0	20	68.0	75	167.0	250	482	425	797	1500	2732
−14	6.8	21	69.8	80	176.0	255	491	430	806	1550	2822
−13	8.6	22	71.6	85	185.0	260	500	435	815	1600	2912
−12	10.4	23	73.4	90	194.0	265	509	440	824	1650	3002
−11	12.2	24	75.2	95	203.0	270	518	445	833	1700	3092
−10	14.0	25	77.0	100	212.0	275	527	450	842	1750	3182
− 9	15.8	25	78.8	105	221.0	280	536	455	851	1800	3272
− 8	17.6	27	80.6	110	230.0	285	545	460	860	1850	3362
− 7	19.4	28	82.4	115	239.0	290	554	465	869	1900	3452
− 6	21.2	29	84.2	120	248.0	295	563	470	878	1950	3542
− 5	23.0	30	86.0	125	257.0	300	572	475	887	2000	3632
− 4	24.8	31	87.8	130	266.0	305	581	480	896	2050	3722
− 3	26.6	32	89.6	135	275.0	310	590	485	905	2100	3812
− 2	28.4	33	91.4	140	284.0	315	599	490	914	2150	3902
− 1	30.2	34	93.2	145	293.0	320	608	495	923	2200	3992
0	32.0	35	95.0	150	302.0	325	617	500	932	2250	4082
+ 1	33.8	36	96.8	155	311.0	330	626	550	1022	2300	4172
2	35.6	37	98.6	160	320.0	335	635	600	1112	2350	4262
3	37.4	38	100.4	165	329.0	340	644	650	1202	2400	4352
4	39.2	39	102.2	170	338.0	345	653	700	1292	2450	4442

Table of Values for Interpolation in the Above Table

Degrees centigrade	1	2	3	4	5	6	7	8	9
Degrees Fahrenheit	1.8	3.6	5.4	7.2	9.0	10.8	12.6	14.4	16.2

TABLE A.10 Three-phase Motors Full Load Currents*

HP	Induction-type squirrel-cage and wound rotor, amperes					Synchronous type unity power factor,† amperes			
	115-volt	230-volt	460-volt	575-volt	2,300-volt	220-volt	440-volt	550-volt	2,300-volt
½	4	2	1	.8					
¾	5.6	2.8	1.4	1.1					
1	7.2	3.6	1.8	1.4					
1½	10.4	5.2	2.6	2.1					
2	13.6	6.8	3.4	2.7					
3	....	9.6	4.8	3.9					
5	....	15.2	7.6	6.1					
7½	....	22	11	9					
10	....	28	14	11					
15	....	42	21	17					
20	....	54	27	22					
25	....	68	34	27	...	54	27	22	
30	....	80	40	32	...	65	33	26	
40	....	104	52	41	...	86	43	35	
50	....	130	65	52	...	108	54	44	
60	....	154	77	62	16	128	64	51	12
75	....	192	96	77	20	161	81	65	15
100	....	248	124	99	26	211	106	85	20
125	....	312	156	125	31	264	132	106	25
150	....	360	180	144	37	...	158	127	30
200	....	480	240	192	49	...	210	168	40

For full load currents of 208- and 200-volt motors, increase the corresponding 230-volt motor full load current by 10 and 15%, respectively.

* These values of full load current are for motors running at speeds usual for belted motors and motors with normal torque characteristics. Motors built for especially low speeds or high torques may require more running current, and multi-speed motors will have full load current varying with speed, in which case the nameplate current rating shall be used.

† For 90 and 80% PF the above figures shall be multiplied by 1.1 and 1.25, respectively.

The voltages listed are rated motor voltages. Corresponding nominal system voltages are 110 to 120, 220 to 240, 440 to 480, and 550 to 600 volts.

TABLE A.11 Natural Trigonometric Functions

Angle	Sin	Tan	Cot	Cos	Deg
0	0.0000	0.0000	∞	1.0000	90
1	0.0175	0.0175	57.2900	0.9998	89
2	0.0349	0.0349	28.6363	0.9994	88
3	0.0523	0.0524	19.0811	0.9986	87
4	0.0698	0.0699	14.3007	0.9976	86
5	0.0872	0.0875	11.4300	0.9962	85
6	0.1045	0.1051	9.5144	0.9945	84
7	0.1219	0.1228	8.1443	0.9925	83
8	0.1392	0.1405	7.1154	0.9903	82
9	0.1564	0.1584	6.3138	0.9877	81
10	0.1736	0.1763	5.6713	0.9848	80
11	0.1908	0.1944	5.1446	0.9816	79
12	0.2079	0.2126	4.7046	0.9781	78
13	0.2250	0.2309	4.3315	0.9744	77
14	0.2419	0.2493	4.0108	0.9703	76
15	0.2588	0.2679	3.7321	0.9659	75
16	0.2756	0.2867	3.4874	0.9613	74
17	0.2924	0.3057	3.2709	0.9563	73
18	0.3090	0.3249	3.0777	0.9511	72
19	0.3256	0.3443	2.9042	0.9455	71
20	0.3420	0.3640	2.7475	0.9397	70
21	0.3584	0.3839	2.6051	0.9336	69
22	0.3746	0.4040	2.4751	0.9272	68
23	0.3907	0.4245	2.3559	0.9205	67
24	0.4067	0.4452	2.2460	0.9135	66
25	0.4226	0.4663	2.1445	0.9063	65
26	0.4384	0.4877	2.0503	0.8988	64
27	0.4540	0.5095	1.9626	0.8910	63
28	0.4695	0.5317	1.8807	0.8829	62
29	0.4848	0.5543	1.8040	0.8746	61
30	0.5000	0.5774	1.7321	0.8660	60
31	0.5150	0.6009	1.6643	0.8572	59
32	0.5299	0.6249	1.6003	0.8480	58
33	0.5446	0.6494	1.5399	0.8387	57
34	0.5592	0.6745	1.4826	0.8290	56
35	0.5736	0.7002	1.4281	0.8192	55
36	0.5878	0.7265	1.3764	0.8090	54
37	0.6018	0.7536	1.3270	0.7986	53
38	0.6157	0.7813	1.2799	0.7880	52
39	0.6293	0.8098	1.2349	0.7771	51
40	0.6428	0.8391	1.1918	0.7660	50
41	0.6561	0.8693	1.1504	0.7547	49
42	0.6691	0.9004	1.1106	0.7431	48
43	0.6820	0.9325	1.0724	0.7314	47
44	0.6947	0.9657	1.0355	0.7193	46
45	0.7071	1.0000	1.0000	0.7071	45
Deg	Cos	Cot	Tan	Sin	Angle

TABLE A.12 American Standard Device Function Numbers

1. Master element
2. Time-delay starting, or closing, relay
3. Checking, or interlocking, relay
4. Master contactor
5. Stopping device
6. Starting circuit breaker
7. Anode circuit breaker
8. Control power disconnecting device
9. Reversing device
10. Unit sequence switch
11. Reserved for future application
12. Overspeed device
13. Synchronous-speed device
14. Underspeed device
15. Speed, or frequency, matching device
16. Reserved for future application
17. Shunting, or discharge, switch
18. Accelerating, or decelerating, device
19. Starting-to-running transition contactor
20. Electrically operated valve
21. Distance relay
22. Equalizer circuit breaker
23. Temperature control device
24. Reserved for future application
25. Synchronizing, or synchronism-check, device
26. Apparatus thermal device
27. Undervoltage relay
28. Flame detector
29. Isolating contactor
30. Annunciator relay
31. Separate excitation device
32. Directional power relay
33. Position switch
34. Motor-operated sequence switch
35. Brush-operating, or slip-ring short-circuiting, device
36. Polarity device
37. Undercurrent, or underpower, relay
38. Bearing protective device
39. Mechanical condition monitor
40. Field relay
41. Field circuit breaker
42. Running circuit breaker
43. Manual transfer, or selector, device
44. Unit sequence starting relay
45. Atmospheric condition monitor
46. Reverse-phase, or phase-balance, current relay
47. Phase-sequence voltage relay
48. Incomplete sequence relay
49. Machine, or transformer, thermal relay
50. Instantaneous overcurrent, or rate-of-rise, relay
51. A-c time overcurrent relay
52. A-c circuit breaker
53. Exciter, or dc generator, relay
54. High-speed dc circuit breaker
55. Power-factor relay
56. Field application relay
57. Short-circuiting, or grounding, device
58. Power rectifier misfire relay
59. Overvoltage relay
60. Voltage balance relay
61. Current balance relay
62. Time-delay stopping, or opening, relay
63. Liquid or gas pressure, level, or flow relay
64. Ground protective relay
65. Governor
66. Notching, or jogging, device
67. A-c directional overcurrent relay
68. Blocking relay
69. Permissive control device
70. Electrically operated rheostat
71. Liquid or gas level relay
72. Dc circuit breaker
73. Load-resistor contactor
74. Alarm relay
75. Position changing mechanism
76. Dc overcurrent relay
77. Pulse transmitter
78. Phase angle measuring, or out-of-step protective, relay
79. Ac reclosing relay
80. Liquid or gas flow relay
81. Frequency relay
82. Dc reclosing relay
83. Automatic selective control, or transfer, relay
84. Operating mechanism
85. Carrier, or pilot-wire, receiver relay
86. Locking-out relay
87. Differential protective relay
88. Auxiliary motor, or motor generator
89. Line switch
90. Regulating device
91. Voltage directional relay
92. Voltage and power directional relay
93. Field changing contactor
94. Tripping, or trip-free, relay
95. ⎫
96. ⎬ Used only for specific applications on individual installations where none of the assigned numbered functions from 1 to 94 are suitable
97. ⎬
98. ⎬
99. ⎭

NOTE: Alternate names such as *relay*, *contactor*, *circuit breaker*, *switch*, or *device* may be used for any function where applicable.
NOTE: Suffix letters are used with device function numbers for various purposes; for instance, suffix N is generally used if the device is connected in the secondary neutral of current transformers, and suffixes X, Y, and Z are used to denote separate auxiliary devices.

TABLE A.13 Selected Symbols

Symbol	Description
Resistor	Resistor
Tapped resistor	Tapped resistor
Adjustable resistor	Adjustable resistor
Variable resistor	Variable resistor
Nonlinear resistor	Nonlinear resistor
Capacitor (If polarized)	Capacitor (If polarized)
Adjustable or variable capacitor	Adjustable or variable capacitor
Single-cell battery with long-line positive	Single-cell battery with long-line positive
Multicell battery	Multicell battery
Conductor	Conductor
Conductor Crossing, not connected	Conductor Crossing, not connected
Conductors connected	Conductors connected
Lighting panel; see panel schedule for rating	Lighting panel; see panel schedule for rating
Power panel; see panel schedule for rating	Power panel; see panel schedule for rating
Transformer; note rating alongside, or refer to single line by number, e.g., T_1, T_2, T_3, T_4, etc.	Transformer
Junction box; 4-in. standard octagon box	Junction box
Junction box; specify size	Junction box; specify size
T Condulet	T Condulet
LB Condulet; identify LB, LR, LL	LB Condulet
C Condulet	C Condulet
Conduit union	Conduit union
Conduit seal fitting	Conduit seal fitting

Symbol	Description
Conduit outlet bodies; combination of T and J boxes with threaded hubs and cover	
El junction box; same as above	
Incandescent light	
Mercury-vapor light	
Fluorescent light (size to scale)	
Wall-mounted floodlight; arrow indicates direction	
$\$_1$	Single-pole switch
$\$_2$	Two-pole switch
	120-volt convenience receptacle; 3-pole two-wire
	440-volt welding receptacle
	Special receptacle (note type)
0-300 V	Meter: F (frequency), SY (synchroscope), VA (voltammeter), V (volts), W (watts), A (amps), AH (amphour), VAR (vars), PF (power factor), WH (watthour); indicate range and type
S	Selector switch
START / STOP	STOP/START pushbutton control station
	HAND/OFF/AUTO control station
	Ground well
– – – – –	Conduit concealed
————	Conduit exposed
——G——	Ground wire

TABLE A.13 Selected Symbols (*Continued*)

Symbol	Description
—W—	Wireway
—○	Conduit turning down
—◉	Conduit turning up
(underground conduit bank section drawing)	Underground conduit bank; show section for contents
1" 2" 3" 1" / 1" 4" 1" 2" / 2" 4" 3" 1½"	Underground conduit bank section. Specify concrete encasement and minimum spacing between conduits, e.g., 3 in. and 1½ in., respectively
150/225	Air circuit breaker; show trip and frame size, e.g., 150-amp trip and 225-amp frame
trip/frame	Air circuit breaker, drawout type
(symbol)	Air circuit breaker with thermal overload device
(symbol)	Air circuit breaker with magnetic overload device
OCB	Circuit breaker other than those covered previously; identify by OCB (oil circuit breaker), PCB (power circuit breaker), etc.
trip/frame 2	Combination magnetic motor starter and air circuit breaker; indicate starter size, e.g., (2)
trip/frame 2	Drawout combination motor starter. NOTE: overload device type can be called out in the specification; therefore it need not be shown.
(3) 50:5	Current transformer; specify quantity and ratio
(3) 50:5	Bushing-type current transformer; specify quantity and ratio
(symbol)	Potential transformer; specify quantity
(symbol)	Outdoor metering device
—◻—	Fuse (general)
—◻—	High-voltage primary fuse cutout (dry)
—▭—	High-voltage primary fuse cutout (oil)
—▷◁—	Current limiter
—○ ○—	Lightning arrester (general)
(R*)	Relay coil
(M*)	Motor starter coil
(C*)	Contactor coil; Or listed identification on the drawing
(symbol)	Terminal strip showing three terminals
(symbol)	Inductor
(symbol)	Two-winding transformer
(symbol)	Autotransformer
(symbol)	Three-winding transformer

TABLE A.13 Selected Symbols (*Continued*)

Symbol	Description	Symbol	Description
	Solenoid		Temperature actuated switch; opens on rising temperature
	Single–throw switch		Thermal cutout
	Double–throw switch	(Contact table: 1-2 C, 3-4 A, 5-6 C, 7-8 A)	Master or control switch; X indicates closed; contact arrangement is shown elsewhere
	Switch with horn gap		
	Pushbutton; momentary close		Limit switch, spring return; normally open
	Pushbutton; momentary open		Limit switch, spring return; normally open, held closed
	Pushbutton; momentary open, close		Limit switch, spring return; normally closed
	Two–circuit maintained; open, close		Limit switch, spring return; normally closed, held open
	Three–position selector switch	TDC	Open switch with time delay on closing
	Flow switch; closes on flow increase	TDO	Closed switch with time delay on opening
	Flow switch; opens on flow increase	TDO	Open switch with time delay on opening
	Liquid-level switch; closes on rising level	TDC	Closed switch with time delay on closing
	Liquid-level switch; opens on rising level	—P	Twisted pair
	Pressure–actuated switch; closes on rising pressure		Shielded conductor
	Pressure–actuated switch; opens on rising pressure		Shielded two-conductor with shield grounded
	Temperature–actuated switch; closes on rising temperature		Two-conductor cable
			Grouping or bundle of conductors

TABLE A.13 Selected Symbols (*Continued*)

Symbol	Description
⏚	Ground connection
⤨	Normally closed contact
⊢⊣	Normally open contact
⊢⊣ TC	Time delay on closing
⤨ TO	Time delay on opening

TABLE A.14 Electrical Formulas

1. va = EI—single-phase and $\sqrt{3}\,EI$—three-phase and $3E_N I$—three-phase
2. Watts = va $\times \cos \phi$
3. $\cos \phi$ = power factor = watts/va = R/Z
4. var = $EI \sin \phi$ for single-phase and $\sqrt{3}\,EI \sin \phi$ for three-phase and watts $\times \tan \phi$
5. kw = watts/1,000 and kva = va/1,000 and kvar = var/1,000
6. kva = $\sqrt{(\text{kw})^2 + (\text{kvar})^2}$ and kw = $\sqrt{(\text{kva})^2 - (\text{kvar})^2}$ and kvar = $\sqrt{(\text{kva})^2 - (\text{kw})^2}$
7. Volts drop$_{LN}$ = $I(R \cos \phi + X \sin \phi)$ and volts drop$_{L1L2}$ = $\sqrt{3}\,I(R \cos \phi + X \sin \phi)$
8. % volts drop = kva$_{3\text{-ph}}$ $(R \cos \phi + X \sin \phi)/10$ (kv)$^2_{LL}$
9. kva$_{sc}$ = $100/\%x$ (base kva) or $100/\%\,z$ (base kva)
10. I_{sc} = kva$_{sc}/\sqrt{3}\,E_{LL}$
11. hp = $\sqrt{3}\,EI \times$ EFF $\times$ PF$/746$ and I = hp $\times 746$/PF $\times$ EFF $\times E \times 1.73$
12. Voltage rise % = kvar$_{cap} \times$ (% trans z)/trans kva for shunt capacitors
13. Volts drop = $IR \cos \phi + I(X_L - X_C) \sin \phi$ for series capacitors
14. $X_c = 1/(2\pi fc)$ and $X_L = 2\pi fL$ L = henries C = farads
15. % z = kva$_{base}$ (% z/kva) and utility % z = (kva base/kva$_{sc}$ utility)10 and utility % z = (kva$_{base}/I_{sc} \times$ kv$_{LL} \times 1.73$)10
16. $z = \sqrt{R^2 + (X_L - X_C)^2}$ and $z = \sqrt{R^2 + X_L^2}$
17. Torque in lb-ft = hp $\times 5{,}250$/rpm

TABLE A.15 The Greek Alphabet

A	α	Alpha	N	ν	Nu
B	β	Beta	Ξ	ξ	Xi
Γ	γ	Gamma	O	o	Omicron
Δ	δ	Delta	Π	π	Pi
Σ	ϵ	Epsilon	P	ρ	Rho
Z	ζ	Zeta	Σ	$\sigma\,\varsigma$	Sigma
H	η	Eta	T	τ	Tau
Θ	θ	Theta	Υ	υ	Upsilon
I	ι	Iota	Φ	ϕ	Phi
K	κ	Kappa	X	χ	Chi
Λ	λ	Lambda	Ψ	ψ	Psi
M	μ	Mu	Ω	ω	Omega

TABLE A.16 Electrical and Magnetic Nomenclature

Corresponding Electrical and Magnetic Quantities and Relationships

Electrical		Magnetic	
Electromotive force	E	Magnetomotive force	F
Current	I	Flux	Φ
Resistance	R	Reluctance	S
Potential	V	Potential	Ω
Field strength	$\mathcal{E}$	Field strength	H
Flux density	D	Flux density	B
Permittivity (free space)	ϵ_o	Permeability (free space)	μ_o
Relative permittivity	ϵ_r	Relative permeability	μ_r
Absolute permittivity	$\epsilon_o \epsilon_r$	Absolute permeability	$\mu_o \mu_r$

$$I = E/R \qquad \mathcal{E} = -\frac{\delta V}{\delta r} \qquad \Phi = F/S \qquad H = -\frac{\delta \Omega}{\delta r}$$

$$V = IR \qquad \mathcal{E} = \frac{D}{\epsilon_o \epsilon_r} \qquad \Omega = \Phi S \qquad H = \frac{B}{\mu_o \mu_r}$$

$$\epsilon_o = 8.85 \times 10^{-12} \text{ mks units} \qquad \mu_o = 12.57 \times 10^{-7} \text{ mks units}$$

Electrical and Magnetic Units

Quantity	Symbol	Name of mks or practical unit
Quantity of charge	Q	coulomb
Resistance	R	
Reactance	X	ohm
Impedance	Z	
Current	I	ampere
Potential difference	V	volt
Electromotive force	E	
Conductance	G	
Admittance	Y	mho
Susceptance	B	
Resistivity	ρ	ohm-m
Conductivity	σ	mho/m
Electric field strength	$\mathcal{E}$	volt/m
Flux density (electric)	D	coulomb/sq m
Capacitance	C	farad
Permittivity	ϵ	farad/m
Self-inductance	L	henry
Mutual inductance	M	
Pole strength	m	weber
Magnetic flux	Φ	weber
Flux density (magnetic)	B	weber/sq m
Magnetomotive force	F	ampere-turn
Magnetic potential	Ω	
Magnetic force	H	AT/m
Permeability	μ	henry/m
Reluctance	S	
Energy	W	joule
Power	P	watt

1 gilbert = $10/4\pi$ ampere-turn.

TABLE A.17 Rule-of-thumb Approximations

MOTOR AMPERES

550-volt three-phase	$I \approx \text{hp} \times 1$
440-volt three-phase	$I \approx \text{hp} \times 1.25$
230-volt three-phase	$I \approx \text{hp} \times 2.5$
208-volt three-phase	$I \approx \text{hp} \times 2.65$
220-volt single-phase	$I \approx \text{hp} \times 5$
110-volt single-phase	$I \approx \text{hp} \times 10$
2300-volt three-phase	$I \approx \text{hp} \times 0.25$

LIGHTING

50 footcandles, fluorescent $\approx$ 2 watts/sq ft
50 footcandles, mercury vapor $\approx$ 2 watts/sq ft
50 footcandles, incandescent $\approx$ 5 watts/sq ft
Spacing approximately equal to mounting height

SHORT CIRCUIT

Maximum $I_{sc} \approx I_{trans} \times 20$
Maximum $kva_{sc} \approx kva_{trans} \times 20$
Maximum at any motor control center, approximately 25,000 amp

POWER FACTOR

Induction motors, 1–10 hp	0.8 Lagging
Induction motors, 10 and up	0.9 Lagging
Induction motors, groups	0.6 Lagging
Welder transformer type	0.6 Lagging
Arc furnaces	0.8 Lagging
Induction furnaces	0.6 Lagging
Fluorescent lighting, high power factor	0.95 Lagging

BRANCH CIRCUITS

Maximum loading, 80%
Convenience receptacles rate at 1.5 amp

Index

Index

Abscissa axis, 202
Absorption law, 148, 232
Accuracy, degree of, in short circuits, 109–110
Active power, definition of, 82
Addition of scalar quantities, 188
Aggregators, 191, 193, 194
Airway connection method, 155
   wiring connection, 154, 155
Alternating current, 10
American standard device function numbers, table of, 248
Ammeter reading, 84–85
Ampacities, table of, 236
Analog instruments, 123
Apparent power, 82
Approval of drawing, 6
Asserted proposition, 221
Associative law, 188, 232
Asymmetrical short-circuit current, 120
Autotransformers, 11, 59–61
Auxiliary contacts, 133–134
Auxiliary relays, 133–134

Available short-circuit current, 106, 110, 121
Axiom:
   of De Morgan's law, 230
   of exclusive OR, 230

Base kva, 111
Bolted short circuit, 106
Bonding, definition of, 163
   (*See also* Grounding)
Boole, George, 221
Boolean algebra, 149, 221
Branch circuits, motor, 22
   conductors, 22, 24
Breakers (*see* Circuit breakers)
Bridge circuits, 147–151
Burglar alarm system, 142

Candlepower, 172, 175
   distribution curve, 172

Index

Capacitors:
  calculation of, 98
  case grounding, 98
  disconnecting devices, 97
  on distribution lines, 93, 95–96
  location of, 93
  optional locations, 97
  in parallel, 98
  switched, 92–93
  system capacity release, 91
  at transformers, 96
Cartesian form, 212–218
Cavity ratio, 168
Certified drawings, 52
Checking philosophy, 182
Circuit breakers, 8–10, 102
  application ratings for, table of, 240–241
  branch, 24
  evaluation of, 103
  interrupting capacity (IC), 106
  location of, 11
  main-feeder, 28
  trip setting in, 26–29
  types of, table of, 239
Circuit logic, 125, 138–151
Coefficient of utilization, 167–168
Color code checking, 182
Color coding, wire, 156
Color content of lighting, 168
Commutative law, 188
Complex number of cartesian form, 212
Complexity, degree of, 101
Conclusion in mathematical logic, 150, 227–229
Conductors, 10
  ampacities, table of, 236
  grounding, 161
  identified, 161
  lightning, 162
  motor-feeder, 23
  neutral, 63, 65, 156, 161
  properties of, table of, 238
  secondary, 32–33
  size of, 28, 29, 33
  supply, 91–92
Conduit fill, table of, 237
Connection diagrams, 152–156
Connections, motor, 56
  wiring, 154, 155
    in airway connection method, 155

Connectives, logic, 127, 129, 223
Contracts, 6–7
Control circuits (*see* Instrumentation and control circuits)
Control power transformers, 62
Controller, motor, 22
Coordinate axis:
  abscissa, 202
  horizontal, 202
  ordinate, 202, 204
  vertical, 202
Cross multiplication, 192–195

Definitions, electrical design, 4
Degree of accuracy, 109–110
Degree of complexity, 101
Delta connection grounding, 159
De Morgan's law, 140, 142, 143, 147, 230
  axiom of, 230
Device function numbers, American standard, table of, 248
Diagrams, wiring and connection, 152–156
Difference of scalar quantity, 188
  subtraction, 188, 190
Differential protection, 66
Digital binary-type relays, 103
Digital instrumentation, definition of, 103, 104, 123
Direct-current motors:
  armature winding, 36
  compound-, series-, and shunt-wound, 36
  exciter, 40–41
  field winding, 36, 39
  generators, 10, 51
  rotor-field, 38
Distribution, underground, 19
Distribution system, 12
Distribution transformer, 12
Distributive law, 188
Division of scalar quantities, 190
Division vector, 209, 216, 218
  by zero, 189–190
Drawing checklist, 20
Drawing list, 19
Drawing numbers, 19
Drawings:
  approval of, 6

Index

Drawings (*Cont.*):
  certified, 52
  designated "hold," 16
  revisions of, 5
  signature on, 6, 181
  single-line (*see* Single-line drawing)
Drives, variable-speed, 52

Electrical formulas, table of, 252
Elementary diagram, 152*n*.
Enclosure-classification design, table of, 244
Energy, 8
Equality, 190
  logical, 224–227
Equipment, transmitting electric, 9
Equipment reactances, 114
Evaluation of fuse and circuit breakers, 103
Exciter, 40–41, 49
  approximate capacity of, 50
  conductor size of, 42
  design calculations of, 43
  oversize, 86
  sample calculation of, 50
  size of, 41
Exciter rating, 49
Exclusive OR, 133, 134, 138, 144
  proof of, 230–232
Explosionproof motor, 36
Exponents, 195–198
  fractional, 197–198
Extra high voltage, 11
Extra's, 6, 7

Fault location, 108
  (*See also* Short circuits)
Feeders:
  circuit breaker for, 27, 28
  conductor for, 23, 28, 29
Field, dc rotor, 38–39
Field current, 37
Fixtures, 170, 175
  selection of, 171
  spacing of, 170
Flowsheet, 122
Fluorescent lighting, 167–169
Footcandles, 166, 168, 171
Formulas, electrical, table of, 252

Fractional exponent, 197, 198
Fractions, 190
  complex, parentheses in, 191–192
Frequency of alternating current, 10
Functions:
  of angles greater than 90°, 206
  common, 205
Fuses:
  vs. circuit breakers, 102
  evaluation of, 103
  with fault currents, 22
Future motor installations, 26–28

Generator conductors, 48
Generator output, 47
Generators, 9–10, 44–51
  ac three-phase, 44–45
  dc, 51
  ratings of, 45–46
Glare, 168
Greek alphabet, table of, 252
Ground fault, 158
Ground location, 160
Ground resistance, 164
  test circuit, 165
  testing of, 164, 165
Ground wells, definition of, 164
Grounding, 157–165
  definition of, 157
  enclosure, 162
  equipment, 158, 161
  nonelectric equipment, 162
  portable equipment, 162
  reactance, 160
  solid, 160
  structural steel, 162, 163
  system, 158
Grounding conductor, 156, 161
Grounding source, 163

HAND/OFF/AUTO switch, 131, 135, 136
High-bay lighting, 167, 171
High voltage, extra, 11
Highway connection method, 155–156
  diagram, 155
Hindrance solution, 139–140, 143, 146
"Hold" on drawings, 16
Horizontal axis, 202
Hypotenuse of right triangle, 201

Identified conductor, 161
Impedance:
  ac circuit, 108
  diagram, 115
  equipment, 109
Impedance parameters, 108, 109
Implications, 145, 220, 221, 227–229
  premise of, 150, 228, 229
Incandescent lighting, 167, 169
Inclusive OR, 138
Inductive heating, 48
Inspection for installations, 6
Instrumentation and control circuits, 122–137
  (*See also* Instruments; Relays; Switches)
Instruments:
  analog, 123
  digital, 123
Interrupting capacity, breaker, 106
Interrupting ratings, 114–115

j operator, 212–217
Job file, 20–21

Kilo, 10
Kilovolts, 8
Kilowatts, 10

Law(s):
  absorption, 148, 232
  associative, 188, 232
  commutative, 188
  De Morgan's (*see* De Morgan's law)
  distributive, 188
  of negation, 222–223
Lighting, 166–180
  art of, 166
  calculations of, 167
  contactor, 178
  floodlighting, 178
  high-bay, 167, 171
Lighting design, 166
Lightning conductors, 162
Line-to-ground fault, 158
Line-to-line fault, 158
Logic:
  absorption of, 148
  analysis of, 29
  AND, 29, 224

Logic (*Cont.*):
  circuit, 125, 138–151
  exclusive OR, 144, 230
  hypothesis of, 31
  implication in, 30, 31, 227
  mathematical, 138, 220–233
    application of, 232
    theorems of, 232–233
  negation in, law of, 222–223
  in networks, 146–148
  OR, 223
  proposition in, 30, 221
  redundancy with, 148–149
Logic connectives, 127, 129, 223
Logic variables, 132, 133, 141
Lumens, lamp, 167, 169

Magnetic starter, 35
Maintenance factor, 167
Mathematical logic, 138, 220–233
Mercury vapor, 168, 169
Modus ponens, 229
Momentary rating, 114
Motor control center, 26, 27, 36, 121
Motor-feeder conductors, 10
Motor-starting characteristics, 53
Motors:
  branch-circuit data, table of, 242
  direct-current, 36–37
  disconnect device in, 23
  full-load currents, table of, 246
  future, 25, 27
  multiple, 25
  oversize, 86
  rotor field, dc, 38–39
  shaded pole, 35
  single-phase, 34–36
  squirrel-cage, 23, 31, 32, 38
  synchronous, 37–38
  three-phase, 23, 34
  variable speed drives, 37, 52
  wound-rotor, 31–34, 56
Multiplication of scalar quantities, 190

National Electrical Code, 3–4
Negated variable, 140–142
Negation, law of, 222–223
Networks, 146, 150
Neutral conductors, 65, 156, 161
Notes, general and sheet, 5
Numbering, schematic, 152–153

Index

Operator j, 212–217
Ordinate axis, 202, 204
Output of ac generators, 46

P&ID (process and instrumentation diagram) sheet, 123
Parameters:
 impedance, 108
 input, 100
Parentheses in complex fraction, 191–192
Percentage values, 109
Permissive solution, 139
Permits for installations, 6
Permittances, 146
Phase angle, 82
Photometric data, 172, 178
Plot plan, 17–19
Point-to-point calculation, 175
Polarity, 68
 additive, 68
 subtractive, 68
Power, 46–47, 81–98
 active, 82
 addition, 84
 apparent, 82, 84
 measurement of, 81
 reactive, 46, 83
 true (real), 46, 82–83, 90
Power factor, 47, 90
 calculations of mixed loads, 93–95
 correction of, 87–89
 plant, 47, 89
 unity, and lagging, 86
Powers of ten, 199
Premise of implication, 150, 228, 229
Product, definition of, 188
Program construction, 126
Project conception, 122
Proof of exclusive OR, 230–232
Proposition, valence of, 143, 144, 183, 221, 228
Protective-device coordination, 101–102
Pythagorean theorem, 204

Quadrants, 205–207, 209, 211
Quantities:
 scalar, 188, 190
 unknown, 190–192

Radical, the, 197–198
 index or order of, 197
Reactances, 108
 equipment, 114
Real numbers, 198
Real (true) power, 46, 82–83, 90
Recloser, definition of, 103
Relays:
 analog-type, 103
 auxiliary, 133–134
 control, 124
 digital binary-type, 103–104
 protective, 8, 123, 124
Revisions of drawing, 5
Rewrites, 127
RFI (radio frequency interference), 163
Rheostats, 36, 51
Rotating equipment, 111
Rotors, 31–34, 56
Rule-of-thumb, 169
 table of, 254

Scalar quantity, 187, 190
Schematic diagram, 152
Schematic numbering, 152–153
Secondary conductors, 32–33
Secondary current, 33
Sectionalizer, definition of, 103
Selector switches, 125
Short-circuit current, 107, 119
Short-circuit magnitude, 119
Short-circuit ratio, 41, 43, 49
Short circuits, 105–121
 asymmetrical, 120
 bolted, 106
 contribution, 114
  circuit dynamics, 114
 degree of accuracy, 109
 study, 107
 symmetrical, 120
Signature on drawing, 6, 181
Single-line drawing, 15–17, 99
 definition of, 15
 for pump house, 26–27
Specifications, job, 7
Squirrel-cage motors, 34, 38
Standard for design procedures, 4
Star/delta conversion, 118
Star/delta grounding, 159
STOP/START circuit, 135–136, 152, 153
Study layout, 101

Index

Substation, 13
Subtraction of scalar quantities, 190
Supply conductors, 91–92
Switches:
    control, 132, 133
    limit, 126
    selector, 125
Symbols, 4
    tables of, 249–253
System, electric, 8
    analysis, 81
System capacity release, 91–92
System study, 99–104
    complexity of, 101
    reasons for, 100
    scope of, 99
    single-line, 99
    study layout, 101

Tangent of angle, 205
Temperature conversion, table of, 245
Theorem(s):
    definition of, 231–232
    of mathematical logic, 232–233
    Pythagorean, 204
Transformer connections, 68
Transformer data, table of, 243
Transformer overcurrent protection, 66
    overcurrent device, 66
Transformers, 8, 57–80
    auto- (*see* Autotransformers)
    constant-current, 65
    control-power, 62
    current in, 58
    differential protection, 67
    distribution, 12–13, 62
    grounding, 63, 65, 159–160
    impedance Z of, 59
    isolating, 62
    polarity, 68
    potential, 62
    power, 61
    principles of, 57–58
    selection of, 58–59
    single-phase, 13, 61
    stepdown, 12
    stepup, 10–11
    three-phase, table of, 243

Transformers (*Cont.*):
    turns ratio, 57
    zigzag, 63, 65, 159
Transmission line, 11–12
Transmitting electric equipment, 9
Triangles, 85, 176
    right, 201
Trigonometric functions, 205, 247
Trigonometry in electrical design, 201–208
Trip setting, 26–29

Underground distribution, 19
Undervoltage protection, 136
Undervoltage release, 36, 131, 136
Unknown quantities, 190–192
Utilization, coefficient of, 167–168

Vacuity rule, 226, 233
Valence of proposition (*see* Proposition, valence of)
Var, 81–88
Variable-speed drives, 51, 52
Variables, 140, 141
Vector manipulation, 209, 210, 217–219
Vector operations, 209, 216–217
Vectors:
    division (*see* Division vector)
    radius, 202
    representation of, 209
    transposing of, 214
Vertical axis, 202
Voltage, extra high, 11
Voltage drop, 24
Voltage regulation, 95
Voltage rise, 96

Wire numbering, 153
Wiring calculations, motors and exciter, 22–43
    (*See also* Exciter; Motors)
Wiring diagrams, 152–156
Wound-rotor motors, 31–34, 56

Zigzag grounding transformers, 63, 65, 159